T0224038

Lecture Notes in Networks and Systems

Volume 31

Series editor

Janusz Kacprzyk, Polish Academy of Sciences, Warsaw, Poland
e-mail: kacprzyk@ibspan.waw.pl

The series "Lecture Notes in Networks and Systems" publishes the latest developments in Networks and Systems—quickly, informally and with high quality. Original research reported in proceedings and post-proceedings represents the core of LNNS.

Volumes published in LNNS embrace all aspects and subfields of, as well as new challenges in, Networks and Systems.

The series contains proceedings and edited volumes in systems and networks, spanning the areas of Cyber-Physical Systems, Autonomous Systems, Sensor Networks, Control Systems, Energy Systems, Automotive Systems, Biological Systems, Vehicular Networking and Connected Vehicles, Aerospace Systems, Automation, Manufacturing, Smart Grids, Nonlinear Systems, Power Systems, Robotics, Social Systems, Economic Systems and other. Of particular value to both the contributors and the readership are the short publication timeframe and the world-wide distribution and exposure which enable both a wide and rapid dissemination of research output.

The series covers the theory, applications, and perspectives on the state of the art and future developments relevant to systems and networks, decision making, control, complex processes and related areas, as embedded in the fields of interdisciplinary and applied sciences, engineering, computer science, physics, economics, social, and life sciences, as well as the paradigms and methodologies behind them.

More information about this series at http://www.springer.com/series/15179

Hiren Kumar Deva Sarma
Samarjeet Borah · Nitul Dutta
Editors

Advances in Communication, Cloud, and Big Data

Proceedings of 2nd National Conference on CCB 2016

 Springer

Editors
Hiren Kumar Deva Sarma
Department of Information Technology
Sikkim Manipal University
Gangtok, Sikkim
India

Nitul Dutta
Department of Computer Science
  and Engineering
MEF Group of Institutions
Rajkot, Gujarat
India

Samarjeet Borah
Department of Information Technology
Sikkim Manipal University
Gangtok, Sikkim
India

ISSN 2367-3370   ISSN 2367-3389   (electronic)
Lecture Notes in Networks and Systems
ISBN 978-981-13-4270-7   ISBN 978-981-10-8911-4   (eBook)
https://doi.org/10.1007/978-981-10-8911-4

Library of Congress Control Number: 2018936630

Printed on acid-free paper

This Springer imprint is published by the registered company Springer Nature Singapore Pte Ltd.
part of Springer Nature
The registered company address is: 152 Beach Road, #21-01/04 Gateway East, Singapore 189721,
Singapore

Preface

Communication has been one of the key aspects of the society since time immemorial. In the last decade, human society has witnessed tremendous changes in communication paradigms based on electronics engineering, computer science and information technology.

Cloud computing is another recent technology which has invaded into modern society. In the present scenario, it is seen that information technology industries heavily rely on cloud technology. Majority of the IT applications exploit cloud technology to a great extent.

Big data analytics is yet another point of focus of the researchers in recent time. The amount of data being generated in every field, every day and also floated in Internet is really huge. It is important to analyse this huge volume of data in order to benefit from it. Therefore, there is need of more sophisticated and smart computing techniques for this purpose.

The purpose of this book is to include articles depicting research results and latest trends in the areas of communication, cloud, big data analytics and some related applications of computing technologies.

The National Conference on Communication, Cloud and Big Data (CCB) 2016, which is second in its series, was organized by the Department of Information Technology at Sikkim Manipal Institute of Technology (SMIT), Sikkim, during 10–11 November 2016. As the title of the conference suggests, the focal points of the conference are communication technologies, cloud computing technologies and big data processing technologies. The aim of CCB 2016 was to converge academicians, researchers, engineers, practitioners and motivated students for exchange of ideas and to discuss recent technological developments in the broad areas of communication, cloud computing and big data analytics. Healthy discussions and sharing of information and experiences took place in these directions during the conference.

The conference received warm response from researchers across the country. Submitted research papers went through review process, and finally, 26 papers were selected for presentation in the conference. The papers received were from different research fields like cloud computing, big data analytics, communication technologies and some applications of computing technologies. Eighteen selected papers of this national conference have been included in this book. Hopefully, readers will be benefited from this book.

We are thankful to all the contributing authors of CCB 2016 and of this book. We thank Sikkim Manipal Institute of Technology for extending all support towards organizing CCB 2016. We are highly grateful to Prof. Janusz Kacprzyk for his encouragement throughout this journey. Last but not least, we thank Springer Nature for supporting us all the way in the process of publication of this book.

Gangtok, India   Dr. Hiren Kumar Deva Sarma (Editor)
Gangtok, India   Dr. Samarjeet Borah (Co-editor)
Rajkot, India   Dr. Nitul Dutta (Co-editor)

Contents

Editors and Contributors

About the Editors

Dr. Hiren Kumar Deva Sarma (Editor) is currently Professor in the Department of Information Technology, Sikkim Manipal Institute of Technology, Sikkim. He received his B.E. degree in Mechanical Engineering from Assam Engineering College in 1998. He completed his Master of Technology in Information Technology from Tezpur University in 2000. He received his Ph.D. degree in 2013 from the Department of Computer Science and Engineering, Jadavpur University. He is the recipient of Young Scientist Award from International Union of Radio Science (URSI) in the XVIII General Assembly 2005, held at New Delhi, India. He has published more than 70 research papers in different international journals, refereed international and national conferences of repute. His current research interests are wireless sensor networks, mobility management in IPv6-based network, cognitive radio networks, network security, robotics, distributed computing and big data analytics.

Dr. Samarjeet Borah (Co-editor) is currently Professor in the Department of Computer Applications, Sikkim Manipal University (SMU), Sikkim, India. He received his Ph.D. degree in Engineering from Sikkim Manipal University (SMU) and M.Tech. degree from Tezpur University, India, respectively. He is involved with three funded projects in the capacity of Principal Investigator/Co-principal Investigator. The projects are sponsored by AICTE (Government of India), DST-CSRI (Government of India) and Dr. TMA Pai Endowment Fund, out of which one is completed and two are undergoing. He is associated with IEEE, ACM (CSTA), IAENG and IACSIT.

He has organized various events in SMU. Some of these events include ISRO-Sponsored Training Programme on Remote Sensing & GIS, National Conference cum Workshop on Bioinformatics and Computational Biology (NCWBCB 2014), North Eastern Regional Workshop on Natural Language Processing (NER-WNLP 2014), First International Conference on Computing & Communication (IC3-2016), First IEEE International Conference on Advanced Computational & Communication Paradigms (ICACCP 2017). He is involved with various journals of repute in the capacity of Editor/Guest Editor such as *International Journal of Synthetic Emotions (IJSE), International Journal of Healthcare Information Systems and Informatics (IJHISI), International Journal of Grid and High Performance Computing (IJGHPC), International Journal of Image Mining (IJIM), International Journal of Virtual Communities and Social Networking (IJVCSN), Informatica, Journal of Intelligent Systems* and *International Journal of Internet Protocol Technology.* He is also associated with some edited books of IGI Global Publication and Elsevier Inc.

 **Dr. Nitul Dutta (Co-editor)** is currently Professor in the Department of Computer Science and Engineering, Marwadi University, Rajkot, Gujarat. He received his B.E. degree in Computer Science and Engineering from Jorhat Engineering College, Assam (1995), and M.Tech. degree in Information Technology from Tezpur University, Assam (2002). He completed his Ph.D. degree in Engineering in the field of mobile IPv6 from Jadavpur University (2012). He has published more than 30 research papers in various journals and conferences of repute. His research interests are mobile IPv6, cognitive radio networks and distributed computing.

Contributors

Rabindranath Bera Department of Electronics and Communication Engineering, Sikkim Manipal Institute of Technology, Majitar, Sikkim, India

Soumyasree Bera Department of Electronics and Communication Engineering, Sikkim Manipal Institute of Technology, Majitar, Sikkim, India

Debjani Bhowmik Department of Computer Science and Engineering, The ICFAI University, Agartala, Tripura, India

S. R. Biradar Information Science and Engineering, SDMCET, Dharwad, India

Debarshita Biswas Department of Computer Science and Engineering, The ICFAI University, Agartala, Tripura, India

Chandralika Chakraborty Department of Information Technology, Sikkim Manipal Institute of Technology, Sikkim Manipal University, Majitar, Sikkim, India

Tejbanta Singh Chingtham Computer Science and Engineering Department, Sikkim Manipal Institute of Technology, Majitar, Sikkim, India

Vipin Choudhary Department of Electronics and Communication Engineering, ASET, Amity University Uttar Pradesh, Noida, Uttar Pradesh, India

Dependra Dhakal Computer Science and Engineering, Sikkim Manipal Institute of Technology, Rangpo, Sikkim, India

Dushyanta Dutta Kaziranga University, Jorhat, Assam, India

Kiran Gautam Sikkim Manipal Institute of Technology, Majitar, India

Aastha Gupta Department of Electronics and Communication Engineering, ASET, Amity University Uttar Pradesh, Noida, Uttar Pradesh, India

Shirshak Gurung Department of IT, Sikkim Manipal Institute of Technology, Majitar, Sikkim, India

Damodar S. Hotkar Computer Science and Engineering, VDRIT, Haliyal, India

Shubham Kapoor Department of Electronics and Communication Engineering, Sikkim Manipal Institute of Technology, Majitar, Sikkim, India

Ferdousi Khatun Computer Science and Engineering Department, Sikkim Manipal Institute of Technology, Majitar, Sikkim, India

Uttam Khawas Sikkim Manipal Institute of Technology, Majitar, India

Sanjay Kumar Oxford Brookes University, Oxford, UK

Moirangthem Goldie Meitei Computer Science and Engineering Department, Sikkim Manipal Institute of Technology, Majitar, Sikkim, India

Anju Mishra Amity University, Noida, Uttar Pradesh, India

Dhruba Ningombam Department of Computer Science and Technology, Sikkim Manipal Institute of Technology, Majhitar, Rangpo, East Sikkim, India

Prativa Rai Department of Computer Science and Engineering, Sikkim Manipal Institute of Technology, Majitar, Sikkim, India

Prerna Rai Department of Computer Science and Technology, Sikkim Manipal Institute of Technology, Majhitar, Rangpo, East Sikkim, India

Riwaz Rai Department of Computer Science and Engineering, Sikkim Manipal Institute of Technology, Majitar, Sikkim, India

Sandesh Rai Computer Science and Engineering, Sikkim Manipal Institute of Technology, Rangpo, Sikkim, India

Priya Ranjan Department of Electrical and Electronics Engineering, ASET, Amity University Uttar Pradesh, Noida, Uttar Pradesh, India

Daniel Ronnow Department of Electronics, Mathematics and Natural Sciences, University of Gavle, Gavle, Sweden

Minakshi Roy Department of Computer Science and Engineering, Sikkim Manipal Institute of Technology, Majitar, Sikkim, India

Shyera Roy Department of Computer Science and Engineering, The ICFAI University, Agartala, Tripura, India

Suchismita Roy Department of Computer Science and Engineering, The ICFAI University, Agartala, Tripura, India

Biswaraj Sen Computer Science and Engineering Department, Sikkim Manipal Institute of Technology, Majitar, Sikkim, India

Kalpana Sharma Computer Science and Engineering, Sikkim Manipal Institute of Technology, Rangpo, Sikkim, India

Pratikshya Sharma Computer Science and Engineering Department, Sikkim Manipal Institute of Technology, Majitar, Sikkim, India

Shanu Sharma Department of CSE, ASET, Amity University, Noida, Uttar Pradesh, India

Additi Mrinal Singh Department of Electronics and Communication Engineering, Sikkim Manipal Institute of Technology, Majitar, Sikkim, India

Tanuja Subba Computer Science and Engineering Department, Sikkim Manipal Institute of Technology, Majitar, Sikkim, India

Samarendra Nath Sur Department of Electronics and Communication Engineering, Sikkim Manipal Institute of Technology, Majitar, Sikkim, India

Pran Hari Talukdar Kaziranga University, Koraikhuwa, Assam, India

Praveen Mukhia Titimus Department of Computer Science and Application, St. Joseph's College, Darjeeling, India

Malay Ranjan Tripathy Department of Electronics and Communication Engineering, ASET, Amity University Uttar Pradesh, Noida, Uttar Pradesh, India; Department of Electronics, Mathematics and Natural Sciences, University of Gavle, Gavle, Sweden

Amit Ujlayan School of Vocational Studies and Applied Sciences, Gautam Buddha University, Greater Noida, India

Survey on Energy-Efficient Routing Protocols in Wireless Sensor Networks Using Game Theory

Riwaz Rai and Prativa Rai

Abstract Wireless Sensor Networks (WSNs) are made up of small low-power nodes that are used in different areas like environment monitoring and several military and civilian applications. But due to its small size and limited energy source, energy efficiency is its main area of concern, and many methods have been developed to improve its network lifetime. Game Theory is being used in WSNs to improve the energy efficiency of a network and its lifetime. Game Theory is suitable for such problems as it can be used in node or network level to encourage the decision-making capabilities of WSNs. This survey paper focuses on different types of clustering protocols designed in WSNs using Game Theory to combat the problem of energy efficiency. In particular, we address the approaches by which Game Theory has been used in WSNs to improve its network lifetime including the games used in each protocol.

Keywords Wireless sensor networks · Game Theory · Energy efficiency
Clustering

1 Introduction

Wireless Sensor Networks (WSNs) incorporate a large number of communication nodes with restricted sensing, processing, and computational capabilities [1]. WSNs are being used in various applications in real life. WSNs have uses ranging from essential social matters such as monitoring of environment and habitat, traffic global position system (GPS), medical emergency, and health services to economic matters like production control and structure monitoring [2]. But there is always a

R. Rai (✉) · P. Rai
Department of Computer Science and Engineering,
Sikkim Manipal Institute of Technology, Majitar, Sikkim, India
e-mail: rewzrai04@gmail.com

P. Rai
e-mail: raiprativa@gmail.com

© Springer Nature Singapore Pte Ltd. 2019
H. K. D. Sarma et al. (eds.), *Advances in Communication,
Cloud, and Big Data*, Lecture Notes in Networks and Systems 31,
https://doi.org/10.1007/978-981-10-8911-4_1

resource constraint due to the nature and property of WSNs. There is always unreliability due to its wireless nature, and there are many energy-related deficiencies due to its power limitation. There are many challenges to make the network efficient to prolong network lifetime along with its own applications due to its constraints.

Routing involves finding the optimum way to transmit data from source to destination nodes in the network. Due to its small nodes, network lifetime is one of its most important factors. Therefore, routing always involves power management in its forefront while designing protocols. Many routing protocols use clustering or hierarchical approach to minimize the consumption of energy during data transmission. Clustering involves creating a cluster or group of clusters to transmit data from source to destination nodes in the network.

Game Theory is used in various walks of life in different applications in different sciences. It has also been used in WSNs. It is used for different aspects in WSNs like mitigating selfish nodes, providing games to increase efficiency, designing routing protocols, security protocols. We can find many works have been carried out using Game Theory to improve different aspects in WSNs. Game Theory came into existence as a branch of economics. It is a mathematical model used to examine as well as predict the actions of rational and selfish individuals.

This paper studies existing proposed game models used for designing clustering algorithm with energy efficiency as the main concern which further leads to energy-efficient routing in WSN.

2 Game Theory

Game Theory was firstly used in economics to make decisions in uncertain conditions. It provides mathematical techniques for analyzing the situations and predicting the future based on the decision taken by individual player. A game is a tuple <N, S, U>, where

N: set of players,
S: a set of actions/strategies for each player and
U: utility or payoff function.

Each player has certain strategy with which he plays the game, and it is defined in set S. Each strategy will consist of some action plan which covers possible situations that can come up in a game. A utility in a game is the players' incentive for playing that strategy. It describes the players' preference and assigns some payoff for each strategy, and the payoff with a larger value is the one that is favored. A Nash Equilibrium is a solution in a game such that the actions of the players do not change even if it knows the strategy of the other players as it does not improve its utility.

2.1 Types of Games

There are different forms of games, and its classification is shown in Fig. 1. A brief description of each of the types is given below:

Cooperative and Non-cooperative Games: The game in which each player knows the strategies of other players and selects the strategy that favors all the players is known as cooperative games, whereas non-cooperative games are games in which no one cooperates with each other and every player is trying to maximize their own profit.

Normal Form and Extensive Form Game: Normal form games are games in which the payoffs as well as strategies used in a game are shown in a tabular format. It can be used to find strategies that are dominated as well as in Nash Equilibrium. The extensive form game is a game where the description of game is done in a decision tree form. It helps in events decided by chance.

Simultaneous-Move and Sequential-Move Games: Simultaneous-move games are games in which the players adopt the strategy simultaneously. Each player is unaware about the strategy of another player, whereas in sequential-move games consist of games where players come to know strategy of earlier players.

Zero-Sum and Nonzero-Sum Games: Zero-sum games are games in which one player's gain amounts to another player's loss, and hence, the sum of outcomes is always zero, whereas in nonzero-sum games, the sum of outcomes is not zero.

Symmetric and Asymmetric Games: Symmetric games are games in which the strategy adopted by all players is same. Here, the payoffs depend on the strategy of the game. Asymmetric games are games where players adopt different strategies. Here, the payoffs depend on the player.

Most games in real life are non-cooperative ones where each node only thinks about itself and its network lifetime. There are also cooperative ones where nodes agree with each other to increase payoffs. Many literature have shown the use of non-cooperative games in WSNs where energy efficiency becomes utmost importance. Thus in such situation, nodes refuse to waste extra energy and conserve its

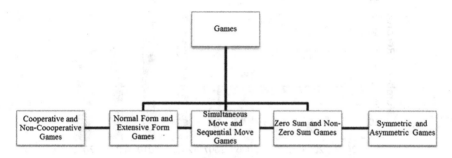

Fig. 1 Types of games

energy by not participating in the process. So the selfish nodes are incentivized by offering bigger payoffs. Also in some cases, selfish nodes are also punished as defected nodes are doubly punished so as to discourage selfish nodes from defecting.

2.2 Games Used for Energy Efficiency in WSN

WSNs are made up of sensors with limited energy supply so the sensors ought to be able to manage energy efficiently while also minimizing its utility and completing its assigned work by communicating in the network. There are different games and approaches taken by the papers in this survey to improve its energy efficiency compared to previous protocols. There are different games also used in these papers to achieve required results.

From [3], it is seen that the following Game Theory methods have been used to formulate games in WSN:

- (i)   Cooperative and non-cooperative game
- (ii) Repeated game
- (iii) Coalitional game
- (iv) Evolutionary game
- (v)   Gur game
- (vi) Bargaining game
- (vii) Bayesian game
- (viii) Transferable and non-transferable utility game
- (ix) Zero and nonzero games
- (x)   Ping-Pong game
- (xi) Jamming game

There have been many research papers in WSN related to Game Theory. Figure 2 shows the distribution of the publications year-wise.

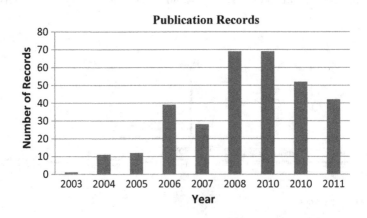

Fig. 2 No. of publications based on Game Theory for WSN [3]

Non-cooperative game is the game which closely resembles the real-life situation as each node is selfish. Non-cooperative Game Theory inspects the relation among competing players, where each player individually selects its strategy and each player's goal is to increase its utility or reduction of its cost. In the survey, Non-cooperative games have been used to find Nash Equilibrium under incomplete information and in one where every node communicates with each other. In another non-cooperative game, a Subgame Perfect Nash Equilibrium is used to select Cluster-Heads. Cooperative game is one in which coalitions are formed by grouping of players, and these players try to strengthen their game position by finding a coalition and act as a single entity by virtue of an agreement. In one, cooperative game is used to find Nash Equilibrium. Coalitional Game Theory is used in another to form coalitions to transmit data. Evolutionary Game Theory (EGT) is another approach. Here, some strategies in the game program the players. From large population, random players are drawn repeatedly and made to play the same pure or mixed strategies. The payoff includes the individual fitness or expected number of surviving offspring. Bayesian game is also used in another approach. In it, we choose the Cluster-Head among rich nodes and poor nodes through a Bayesian game. In it, Bayesian Nash Equilibrium is reached through Bayesian games.

3 Related Works

In this section, we present approaches and protocols used by researchers to improve the energy efficiency in WSNs using Game Theory.

Koltsidas and Pavlidou [4] formulated a clustering mechanism called Clustered Routing for Selfish Sensors (CROSSs) which has the critical aspect as the random rotation of Cluster-Head's role for the goals of energy balancing, based on the rationality and selfishness of the sensor nodes in the network. In this paper, clustering game (CG) is used with a utility function that involves a payoff of 0 to a node if all nodes do not participate and a utility of $v-c$ if it becomes a Cluster-Head and v for a node if it does not become a Cluster-Head but some other node is one (where c is the expense of becoming a CH and v is the gain amount for successful data transmission to Base Station). It is a case of mixed strategies Nash Equilibrium. In this, the players select their strategy randomly succeeding a probability distribution. Thus, the nodes compute the probability of becoming a CH. And randomly some nodes declare themselves as CHs and alert others so that other nodes can send data to the nearest CH. CHs thus aggregate and send the data to the sink. It is then repeated. Zero Probability Rule (ZPR) is also used to set the probability of nodes that have been CHs to zero until every neighbor has been a CH, and then, it switches back it to normal. However, the paper assumes that every sensor communicates with each and every other sensor.

Xie et al. [5] formulated a new algorithm and named it as Localized Game Theoretical Clustering Algorithm (LGCA). In this approach, energy efficiency is improved by using Game Theory (mixed strategy Nash Equilibrium) and a repeated clustering game. The game and utility function is similar to the [4], i.e., Clustering game. It differs though due to the fact that in it, every node plays a clustering game on its own and also joins the games of their neighbors' to get its equilibrium probability. This helps nodes declare themselves as CHs. It also uses Carrier Sense Multiple Access with Collision Avoidance (CSMA/CA) mechanism as avoidance of closeness between two CHs. Its drawback is that it assumed that every node's parameter is the same. Also, various important parameters like node degree, residual energy, and distance to Base Station (BS) are not of concern in this paper. Also, final CHs should have more preference toward powerful nodes instead of any nodes.

Yang et al. [6] proposed a hybrid distributed clustering protocol based on Game Theory called HGTD (Hybrid, Game Theory based and Distributed clustering). This proposed algorithm contains two phases: (i) initialization and (ii) setup phase. Initialization phase lets nodes calculate the distance from the node to the Base Station and also its neighbors. The setup phase consists of tentative CHs selection, final CHs election, and clusters formation. Tentative CHs selection lets each node play various clustering games with itself and its neighbors. It then plays a clustering game on its own and calculates the equilibrium probability to decide by itself whether to be a CH. It then broadcasts and receives messages to know its neighbor tentative CHs. The final CHs are elected to ensure an even distribution of CHs. Any tentative CHs with no neighbor tentative CHs automatically becomes a final CH. For others, it starts an iterative process in which tentative nodes are selected when each node select a CH based on the lowest cost. The nodes become the tentative nodes for next round. Then if the node is in $n$max (i) iteration and also is the tentative CH plus has the lowest cost within a radius R or it does not have any neighboring CHs, it becomes a final CH and it then broadcasts the message within radius R. In cluster formation, the normal node with neighboring final CHs selects the node with the smallest node degree to join the cluster. It also improves on LGCA by assigning more weight toward potential CHs with fewer neighbor CHs and more residual energy to improve it.

Kazemeyni et al. [7] proposed an enhanced group membership protocol for WSNs which selects the best available group the node can have. It uses coalition Game Theory in cooperative games. It uses a proposed power-sensitive AODV routing protocol that detects the cheapest route from node to the leader. In this paper, a coalitional game (N, v) is proposed consisting of N players and utility function or coalition value defined by a characteristic function v: $2N \rightarrow R$ for each players in coalition. In coalition game, a player does not earn any incentives in joining or forming a new coalition. It uses the utility function $\omega(P_j, \delta_j) = \left(\frac{R}{P_j}\right)\left(1 - e^{-0.5\delta j}\right)^L$[7]. When applying ω to a node j, P_j is the power used by j to transfer message and δ_j is the signal-to-interference-and-noise ratio (SINR) for j. Furthermore, R specifies the rate of information transmission in L bit packets in the WSN. The protocol modifies the AODV protocol by having every group leader use the power-sensitive AODV protocol and in turn find the cheapest path for a node j as a possible group member and assess the gain of group membership for j: $g_i(M \cup \{j\}, N) < g_i(M, N \cup \{j\})$). If utility is improved, then the leader sends an Invite message along with utility values both old and new to j. The node j computes the benefit and accepts the Invite message. The Leader accepts and updates its utility.

Zheng et al. [8] used the Game Theory's theorem to analyze routing in WSNs. It also proposes a clustering routing algorithm based on the Bayesian game called Bayesian Game Clustering Routing Algorithm (BGCRA). BGCRA uses separation of nodes into poor and rich nodes. It is different from others as it uses Bayesian game but also it assumes that the network consists of practical heterogeneous WSNs, whereas most protocols assume the homogeneous network which is not true in a real-life situation. In a practical heterogeneous WSN, there may be different types of sensor with different preferences and characteristics and different power levels. In this algorithm, the nodes are divided into rich and poor nodes in regard to remained energy class, and the preferences for both nodes are different for both. But it also needs to be more inclusive in division of nodes other than only rich and poor.

Mishra et al. [9] proposed a Cluster-Heads (CH) selection algorithm named as Game Theory-Based Energy-Efficient Cluster-Head Selection Approach (GECSA) using Subgame Perfect Nash Equilibrium (SPNE) decision of Game Theory. In this paper, a Game G that is non-cooperative is considered. It is assumed that the number of players is n and every cluster is seen as a player in the game, and there are k nodes in each cluster. Every player has q number of strategies denoted as S. The residual energy $E_{residual}$ is fixed as a strategy for each player p_i. Then the action of one player is checked with another player to see the greater one. After getting SPNE, the one is selected as a CH. There are different numbers of strategies for each player in the game, and each player according to his payoffs chooses his best strategy. Every player selects its best strategy amidst all SPNEs denoted as s_{max} and chooses it as a CH in the event of multiple SPNEs.

For CH distribution in every region, there is a statistical allocation of sensor nodes in a finite space. After selecting the CHs, each CH broadcasts announcement packets within a radius suppose $\beta \; r_i$, where β is the system parameter. Every non-CH should receive the packet. From these packets, the non-CH chooses a CH as their own and sends the information. The CH first aggregates the data before forwarding it to the sink node via another CH.

Raja and Dananjayan [10] have proposed a Game Theory based routing protocol which enhances the lifetime of WSN. It uses Cooperative Multiple Input Multiple Output (CMIMO) scheme along with Efficient Energy Consumption Protocol (EECP) proposed by them. In this paper, EECP is used at the beginning to select CHs. It contains (1) CH election stage (2) intracommunication stage, and (3) intercommunication stage. In every round, the above stages are repeated. The first stage consists of election of CH, and the method used is randomized maximum weight selection method. The second stage (intracommunication stage) consists of data collection and aggregation by CH from its cluster members. The third stage (intercommunication stage) consists of forwarding the aggregated data to the sink node through CHs and CNs. Then, CMIMO is used as a routing scheme. It consists of three phases: cluster formation, intra- and intercluster communication.

In CMIMO using maximum weight selection method, the CHs are selected. Then, the selected CH nodes choose cooperative sending nodes and cooperative receiving nodes for CMIMO communication based on the weights in accordance to each nodes energy availability. Nodes with higher weights nearer to the elected CH will be the sending and receiving cooperative group nodes for the cluster. Here, coalitional Game Theory is used for the selection of CNs and selection is done using the basis as the distance and residual energy of the nodes. The coalitional game is modeled as (N, v, V) where N is the set of players (nodes), $\{1, 2..., n\}$, v is the characteristic function based on the network lifetime, V is the partition of N, $V \subseteq N$. It is, however, not suitable for short distances but effective in longer distances. There is also a delay problem.

Tyagi et al. [11] proposed a Bayesian Coalition Game-based optimized clustering in WSNs using the concepts of Learning Automata (LA) and Bayesian Coalition Game (BCG) where Sensor Nodes (SNs) are considered as the players in the game with dynamic threshold-based coalition formation among themselves, i.e., the nodes form coalition based on distance-based thresholds making a partition of the network field. In the process, a reward or penalty is assigned to every player in accordance to the finite number of actions performed. One of the SNs is elected as CH based on the subfield. So in this approach, data can be sent to BS directly or through CHs. An LA-assisted coalition game-based clustering scheme is being used. Node densities are calculated as G_1, G_2 of the calculated subregions.

If $G1 = G2$, LA's action is rewarded with updates to action probability vector. Penalty is also performed according to the above. The computation of ratio of rewards and a penalty is done, and a new coalition is joined if and only if the value is lower than predefined threshold else it continues on in the current coalition. CH is chosen as the node having maximum value of PF. It broadcasts the message to CMs using TDMA schedule.

Tan et al. [12] proposed an algorithm for Cluster-Head selection based on cost sharing game known as Cost Sharing Game-based Clustering (CSGC). It is used to select CHs and fair allocation of cost. It uses a cost sharing game made up of a set N with m agents and a cost function C. It also uses Shapley value and cost shares (ξ) to distribute the total cost equitably.

In this paper in a cluster, a cost sharing game is used with three agents as PCH_0, PCH_1, PCH_2 where one is in center, other has additional energy, and the other is in the boundary of a cluster. So a potential CHs is elected (PCH_0) and other potential CHs are recruited according to above conditions. Now, CMS are recruited. It is done by same conditions as CHs. Only non-CHs are allowed. It is used to relay data in their time slots and sleeps in other time. CHs coalition is used to share cost.

4 Conclusion

This report starts by a brief introduction of WSN, its application, and different aspects of WSN. The limitations of sensor nodes in WSNs are also briefly discussed. Game Theory and its components are introduced, and Nash Equilibrium is also noted. Game Theory is discussed as an approach to combat energy efficiency issues faced by WSNs. The different games used by paper surveyed are discussed in brief. We have discussed the cooperation or non-cooperation of nodes in a network. We have discussed the various ways the papers tried to improve energy efficiency, be it by using clusters and Cluster-Head or mitigating selfish nodes by punishing nodes or rewarding others handsomely for participating. Most papers assume the homogeneity of networks and also that communication between nodes are always there. It also shows the improvement of different protocols used for energy efficiency in subsequent papers.

References

1. Akyildiz IF, Su W, Sankarasubramaniam Y, Cayirci E (2002) Wireless sensor networks: a survey. Comput Netw 38(4):393–422
2. Alemdar H, Ersoy C (2010) Wireless sensor networks for healthcare: a survey. Comput Netw 54(15):2688–2710
3. Shi HY, Wang WL, Kwok NM, Chen SY (2012) Game theory for wireless sensor networks: a survey. Sensors 12(7):9055–9097
4. Koltsidas G, Pavlidou FN (2011) A game theoretical approach to clustering of ad-hoc and sensor networks. Telecommun Syst 47(1–2):81–93
5. Xie D, Sun Q, Zhou Q, Qiu Y, Yuan X (2013). An efficient clustering protocol for wireless sensor networks based on localized game theoretical approach. Int J Distrib Sens Netw
6. Yang L, Lu YZ, Zhong YC, Wu XG, Xing SJ (2016) A hybrid, game theory based, and distributed clustering protocol for wireless sensor networks. Wireless Netw 22(3):1007–1021
7. Kazemeyni F, Johnsen EB, Owe O, Balasingham I (2011) Group selection by nodes in wireless sensor networks using coalitional game theory. In: 16th IEEE international conference on engineering of complex computer systems (ICECCS), IEEE, pp 253–262
8. Zheng G, Liu S, Qi X (2012) Clustering routing algorithm of wireless sensor networks based on Bayesian game. J Syst Eng Electron 23(1):154–159
9. Mishra M, Panigrahi CR, Sarkar JL, Pati B (2015) GECSA: a game theory based energy efficient Cluster-Head selection approach in Wireless Sensor Networks. In: 2015 international conference on man and machine interfacing (MAMI), IEEE, pp 1–5
10. Raja P, Dananjayan P (2015) Game theory based cooperative MIMO routing scheme for lifetime enhancement of WSN. Int J Wireless Inf Netw 22(2):116–125
11. Tyagi S, Tanwar S, Gupta SK, Kumar N, Misra S, Rodrigues JPC, Ullah S (2015) Bayesian coalition game-based optimized clustering in wireless sensor networks. In: 2015 IEEE international conference on communications (ICC), IEEE, pp 3540–3545
12. Tan L, Zhang S, Jin Q (2012) Cooperative cluster head selection based on cost sharing game for energy-efficient wireless sensor networks. J Comput Inf Syst 8(9):3623–3633

Design and Analysis of Optimum APSK Modulation Technique

Shubham Kapoor, Soumyasree Bera and Samarendra Nath Sur

Abstract This paper investigates the design of power and spectrally efficient coded modulations based on amplitude phase shift keying (APSK) modulation with application to satellite broadband communications. This paper presents the error rate analysis of circular APSK modulation with additive white Gaussian noise (AWGN) channel, and an approach to design optimum APSK configuration is being discussed. A circular APSK constellation can have a large number of configurations based on the number of points present in each ring and a total number of rings, which have variable bit error rate based on the configuration. The approach tries to find the optimum APSK configuration and analyzes their performance in Rician and AWGN channels.

Keywords APSK · AWGN · BER · Euclidean · PSK · QAM
Rician

1 Introduction

M-ary star QAM schemes are spectrally efficient modulation schemes which ensure higher data rates without requiring extra bandwidth, which are essential for future generation communication system. It is an amalgamation of amplitude and phase-shift keying, also termed as APSK. In APSK, the signal points appear as phase-shift keying (PSK) signal points distributed over concentric rings.

S. Kapoor (✉) · S. Bera · S. N. Sur
Department of Electronics and Communication Engineering,
Sikkim Manipal Institute of Technology, Majitar, Sikkim, India
e-mail: kpr_shubham@yahoo.com

S. Bera
e-mail: soumyasree.bera@gmail.com

S. N. Sur
e-mail: samar.sur@gmail.com

© Springer Nature Singapore Pte Ltd. 2019
H. K. D. Sarma et al. (eds.), *Advances in Communication,
Cloud, and Big Data*, Lecture Notes in Networks and Systems 31,
https://doi.org/10.1007/978-981-10-8911-4_2

Recent development in satellite communications urges to replace or complement conventional modulation schemes with higher-order M-ary modulation schemes as can be seen in [1]. Rectangular QAM constellations are sub-optimal in the sense as the constellation points for a given energy are not optimally spaced. Due to high PAPR of rectangular and square QAM, they are generally not preferred as in [2]. Today, multilevel modulation schemes in broadband communication systems are a hot research topic. Such kinds of modulation schemes are M-ary phase-shift keying (M-PSK), quadrature amplitude modulation (QAM), and amplitude and phase-shift keying (APSK). Several configurations were proposed [3], and optimization of constellation points has been presented. However, brute-force method produces the optimized solution. The configuration complexity increases when M-ary (number of points) is high. In [3, 4], optimization of the constellation takes into consideration the minimum of inter-ring and intra-ring distances as Euclidean distance and takes the maximum value of the Euclidean distance among different constellations as the optimum configuration. In [2, 5], the analysis of APSK constellation in terms of the equation of circle was presented. In [6], the optimal constellation design is achieved by balancing the SER caused by inter-ring and intra-ring errors. In [7], a theoretical expression for symbol error rate of M-APSK constellation was presented. In [8], the Rician channel was discussed with respect to APSK modulation. The paper tries to present an approach using [3, 5] for optimum constellation design.

2 Mathematical Description

2.1 APSK Constellation

In [3], it was seen that the constellation points are distributed over a point s_i at a distance of r_1 and r_2 from the center, placed at a uniform spacing with a constant phase difference. For a point s_i placed on the ith ring

$$s_i = re^{j\theta_i}$$

and

$$\theta_i = \frac{2\pi}{M}, \tag{1}$$

where r_i = radius of ith ring and M = number of signaling point in one ring. M divides the circle for points to have constant phase difference. For a 4 + 12 ring constellation with 4 points in inner ring at distance r_1 and 12 points at distance r_2, the expression becomes

Fig. 1 Constellation diagram
of 4 + 12 ring from [4]

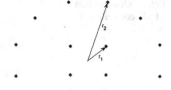

$$S_i = r_1 e^{j\frac{2\pi i}{4}} \quad \text{for } i = 1, 2 \ldots 4 \tag{2}$$

$$S_j = r_2 e^{j\frac{2\pi j}{12}} \quad \text{for } j = 1, 2 \ldots 12 \tag{3}$$

The 16-point constellation for a 2-ring Ring QAM (4 + 12) with SNR = 20 dB is given in Fig. 1.

The expression for three rings with sample point 1, 5, and 10 at a distance r_1, r_2, r_3 from the center is

$$S_i = r_1 e^{j\frac{2\pi i}{4}} \quad \text{for } i = 1, 2 \ldots 4 \tag{4}$$

$$S_k = r_2 e^{j\frac{2\pi k}{8}} \quad \text{for } k = 1, 2 \ldots 8 \tag{5}$$

$$S_l = r_3 e^{j\frac{2\pi l}{16}} \quad \text{for } l = 1, 2, 3 \ldots 16 \tag{6}$$

The 32-point constellation of 3-ring star QAM for transmitted and received signal with SNR = 20 dB is shown in Fig. 2.

Focusing on 16-APSK, let us partition the constellation into points N_1 and N_2 in the inner ring and points in the outer ring with radii and R_1 and R_2, respectively. The minimum distance of the constellation depends on the number of rings, the number of points in each ring $n_1, n_2 \ldots n_R$, the radii $r_1, r_2 \ldots r_{N_r}$, and the phase offset. The constellation geometry clearly indicates that the distances to consider are between points belonging to the same ring or between points in adjacent rings. Figure 3 illustrates the two distances in APSK constellation.

Fig. 2 Constellation diagram
of 1 + 5 + 10 ring from [4]

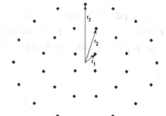

Fig. 3 Optimization of
APSK constellation

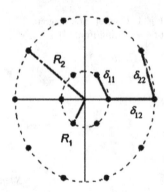

Simple calculations in [4] give the following formula for intra-ring distance (distance between points in same ring)

$$\delta_{11} = 2 \cdot R_1 \sin\left(\frac{\pi}{N_1}\right) \tag{7}$$

and

$$\delta_{22} = 2 \cdot R_2 \sin\left(\frac{\pi}{N_2}\right) \tag{8}$$

For the distance between points in inner and outer rings (inter-ring distance).

For the adjacent rings, calculation is only slightly more complicated and gives the following:

$$\delta_{12} = \sqrt{r_1^2 + r_2^2 - 2r_1r_2\cos\theta} \tag{9}$$

where θ is the minimum relative offset between any pair of points of rings i and $i + 1$, respectively.

The minimum distance of the constellation is given by taking the minimum of all these inter-ring and intra-ring values

$$\delta_{\min} = \min\left(\delta_{\text{ring } i}, \delta_{\text{ring } i,i+1}\right) \tag{10}$$

As per [4], the Euclidean distance can be used to compare different APSK configurations for any number of points. The configuration with the highest Euclidean distance has the lowest bit error rate (BER).

2.2 Symbol Error Rate (SER) Analysis in AWGN Channel

As in [7], SER calculation has been done considering the minimum Euclidean distance. It can be expressed as

$$P_s < (M-1)\sqrt{\frac{d_{min}^2}{2N_0}} \qquad (11)$$

Here, d_{min} = Euclidean distance.

As in [5], the formula of symbol error rate for M-ary circular APSK is

$$P_s \approx \sum_{k=0}^{i} \frac{N_k}{M} \text{erfc}\left(\frac{\widehat{d_i}}{2}\sqrt{\frac{E_s}{N_o}}\right) + \sum_{l=1}^{i} \frac{N_l}{M} \text{erfc}\left(\frac{\Delta\widehat{r_l}}{2}\sqrt{\frac{E_s}{N_o}}\right) \qquad (12)$$

where

$$\widehat{d_i} = \frac{d_i}{\sqrt{E_s}}$$

$$\Delta\widehat{r_l} = \frac{\Delta r_i}{\sqrt{E_s}}$$

where E_s is the energy per symbol.

2.3 Optimization of APSK Constellation

The SER of circular APSK modulation scheme is a function of

1. Number of alphabets in each ring,
2. Intra-ring distance.

For each set of constellation configuration, there is a set optimal inter-ring distances.

$$\frac{dP_s}{d\Delta r} = \frac{\partial P_{s\_d}}{\partial \widehat{d_0}} \frac{\partial \widehat{d_0}}{\partial \Delta r} + \frac{\partial P_{s\_r}}{\partial \widehat{\Delta r}} \frac{\partial \widehat{\Delta r}}{\partial \Delta r} \qquad (13)$$

The optimal solution can be found by solving the differential function.

$$P_{s\_d} = \mathrm{erfc}\left(\sqrt{\frac{E_S}{N_0}}\right)$$

$$P_{s\_r} = \sum_{t=1}^{M} \frac{N_i}{M} \mathrm{erfc}\left(\frac{\Delta\hat{r}}{2}\sqrt{\frac{E_S}{N_0}}\right)$$

(14)

As per the study, optimal SER can be achieved when intra-ring distance $d_0 = \Delta r$; that is, both the inter-ring and intra-ring distances are equal.

To achieve the condition for optimal bit error rate, for any given d_0 of the basic ring.

1. Make the inter-ring distances between each ring equal to d_0
2. Calculate the capacity N_i of ith ring with d_0

From the rule, the radii and the maximum capacity of alphabet set of ith ring could be expressed as

$$r_i = r_o\left(1 + 2 \cdot i \cdot \sin\frac{\pi}{N_0}\right)$$

(15)

$$N_i \le \mathrm{round}\left\{N_o \cdot \left(\frac{r_i}{r_o}\right)\right\}$$

(16)

For any M-ary APSK constellation, the best configurations are the set of N_i where sum $(N_i) = M$ with minimum rings. Table 1 lists the calculated optimum radius with respect to inner radius $r_0 = 1$ using Eq. (18). Table 2 lists the maximum number of points that can be placed in a single ring so as to have equal intra-ring distance using Eq. (19). As the inter-ring and intra-ring distances are equal, the inter-ring distance can be taken as the normalized Euclidean distance.

As per the theoretical analysis, to achieve optimal BER, the two distances, inter-ring and intra-ring distances, are equal. To achieve it, with respect to number

Table 1 Radius of different M-ary constellations from [5]

M-ary on basic ring	d_0	N_0	N_1	N_2	N_3	N_4	N_5	N_6
2	2	2	6	10	14	18	22	26
3	1.73	3	8	13	19	24	29	34
4	1.41	4	10	15	21	27	32	38
5	1.18	5	11	17	23	29	34	40
6	1	6	12	18	24	30	36	42
7	0.87	7	13	19	25	31	37	43
8	0.77	8	14	20	26	32	39	45

Table 2 Capacity of rings at different M-ary constellations from [5]

M-ary on basic ring	d_0	r_0	r_1	r_2	r_3	r_4	r_5	r_6
2	2	1	3	5	7	9	11	13
3	1.73	1	2.73	4.46	6.2	7.93	9.66	11.3
4	1.41	1	2.41	3.83	5.24	6.66	8.07	9.49
5	1.18	1	2.18	3.35	4.53	5.7	6.8	8.05
6	1	1	2	3	4	5	6	7
7	0.87	1	1.87	2.74	3.6	4.47	5.34	6.21
8	0.77	1	1.77	2.53	3.3	4.0	4.83	5.59

of M-ary alphabets in initial ring, the maximum number of points in each ring was calculated for which these conditions were satisfied.

For any configuration of APSK, the radius is taken from Table 1 as per the capacity of each ring in Table 2. Now, the normalized Euclidean distance (inter-ring distance) for any M-ary constellation configuration is also given. The configuration having the highest intra-ring distance for an M-ary configuration is optimum. Also, to any M-ary constellation, having minimum rings is said to be more optimum so as to have lesser power as per [6].

Table 3 gives the best configuration for 16-to 256-point APSK, based on highest inter-ring distance and minimum number of rings.

It may happen that two configurations designed using the above criteria have same inter-ring distance. To differentiate those, we use normalized distance.

The normalized inter-ring distance can be calculated when power is normalized to one. It is a measure of optimum APSK configuration. To calculate it, the first the normalized radius is calculated from the normalized power

$$\frac{N_1 R_1^2 + N_2 R_2^2 + \cdots + N_K R_K^2}{M} = 1 \tag{17}$$

From the values of normalized radius calculated in above, it is multiplied by the intra-ring distance to get the normalized ring radius.

Table 3 Optimal ring configurations based on highest inter-ring distance

No. of points (M)	Ring configuration	Inter-ring distance d_0
16	5 + 11	1.18
32	5 + 11 + 16	1.18
64	7 + 13 + 19 + 25	0.87
128	7 + 13 + 19 + 25 + 31 + 33	0.87

3   Results

3.1   64 APSK Optimization

Using data from Tables 1 and 2, 64 APSK configurations were designed as in Table 4.

Here, 8 + 14 + 20 + 26 + 32 and 10 + 16 + 20 + 24 + 28 + 30 have the same intra-ring distance. In such a case, a new parameter, the value of normalized inter-ring distance is used.

For a 7 + 13 + 19 + 25 ring.

From (11), we have

$$r_i = r_o\left(1 + 2 \cdot i \cdot \sin\frac{\pi}{7}\right) \tag{18}$$

$$r_1 - r_0 = 2 \cdot r_0 \cdot \sin\frac{\pi}{7} \tag{19}$$

$$\frac{d0}{r0} = 2 \ \sin\frac{\pi}{7} = 0.867 \tag{20}$$

$$\text{Mean Power} = \frac{7 * r_0 + 13 * r_1^2 + 19 * r_2^2 + 25 * r_3^2}{64} = 1$$

$$\frac{7 * r0^2 + 13(r0 + d0)^2 + 19(r_0 + 2 * d_0)^2 + 25(r_0 + 3 * d_0)^2}{64} = 1$$

Also, $r_1 - r_0 = d_0$

$$\frac{7 * r0^2 + 13 * r_0\left(1 + \frac{d_0}{r_0}\right)^2 + 19 * r_0\left(1 + 2 * \frac{d_0}{r_0}\right)^2 + 25 * r_0\left(1 + 3 * \frac{d_0}{r_0}\right)^2}{64} = 1$$

From (17)

$$\frac{7 * r0^2 + 13 * r_0(1.867)^2 + 19 * r_0(2.734)^2 + 25 * r_0(3.601)^2}{64} = 1$$

On solving, normalized $r_0 = 0.3513$.

Table 4 Optimum two-ring 64 APSK configurations with highest inter-ring distance

Configuration	d_0	r_1	r_2	r_3	r_4	Normalized d_o
7 + 13 + 19 + 25	0.87	1	1.87	2.74	3.6	0.3054
8 + 14 + 20 + 22	0.77	1	1.77	2.93	3.3	0.30
9 + 15 + 20 + 20	0.68	1	1.68	2.37	3.05	0.2901
4 + 14 + 20 + 26	0.77	1	1.77	2.53	3.3	0.186

Configuration	Normalized d_o
7 + 13 + 19 + 25	0.3054
8 + 14 + 20 + 22	0.30
9 + 15 + 20 + 20	0.2901
4 + 14 + 20 + 26	0.186

Table 5 Normalized distance of 64 APSK configurations

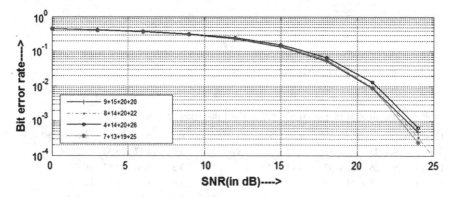

Fig. 4 Comparison of 64 APSK configurations

In (20)

$$\text{Normalized } d_0 = r_0 * 2 * \sin\frac{\pi}{7} = 0.3056$$

The ring with the maximum normalized distance is the one having optimum BER. Table 5 gives the calculated value of normalized distance.

The configurations exhibit lesser variable bit error rate in increasing order of normalized distance, as can be seen in Fig. 4.

3.2 128 APSK Optimization

Using the data from Tables 1 and 2, some 128 APSK configurations were optimized as per table

The BER curve is as per Table 4 in Fig. 5. The 7 + 13 + 19 + 25 + 31 + 33 is the best constellation for 128 points having largest normalized distance (Table 6).

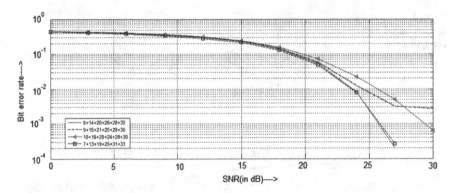

Fig. 5 Comparison of 128 APSK constellations

Table 6 Optimization of 128 APSK with highest inter-ring distance

Configuration	d_0	r_1	r_2	r_3	r_4	r_5	r_6	Normalized d_o
7 + 13 + 19 + 25 + 31 + 33	0.87	1	1.87	2.74	3.6	4.47	5.34	0.3054
8 + 14 + 20 + 26 + 28 + 32	0.77	1	1.77	2.93	3.3	4.06	4.83	0.30
9 + 15 + 21 + 25 + 28 + 30	0.68	1	1.68	2.37	3.05	3.74	4.42	0.2901
10 + 16 + 20 + 24 + 28 + 30	0.77	1	1.62	2.85	3.47	4.09	4.70	0.186

3.3 Response in Rician Channel

As per [8], Rician distribution is given as

$$f(\gamma) = \frac{1+K}{\bar{\gamma}} \exp\left[\frac{-\gamma(1+K)}{\bar{\gamma}} - K\right] X I_0 \left[2\sqrt{\frac{\gamma K(1+K)}{\bar{\gamma}}}\right] \tag{21}$$

where I_0 denotes modified Bessel function of first kind and zero order. $\bar{\gamma}$ denotes average SNR, and γ is SNR. The parameter K gives the extent to which the LOS component is present in signal.

Figures 6 and 7 show the performance of 64 and 128 APSK constellations for different values of k.

From Fig. 6a, b, it is seen that BER is decreasing for both 64 and 128 APSK configurations, with the increase in value of K parameter. A comparison of 64 and 128 APSK configurations was performed as can be seen in Fig. 7.

It is seen that even though BER is decreasing for both the cases with increase in the value of K, it is more in case of 128 APSK configurations owing to more number of points in the same ring.

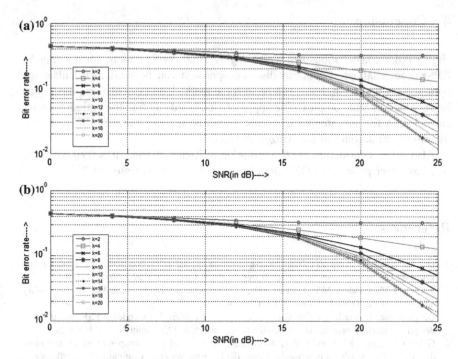

Fig. 6 a BER performance comparison of 64 APSK configurations in Rician channel. **b** BER performance comparison of 128 APSK configurations in Rician channel

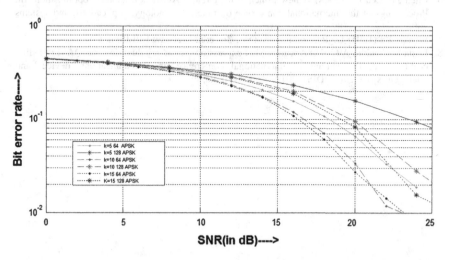

Fig. 7 BER performance comparison of 64 and 128 APSK configurations in Rician channel

4   Conclusion

Higher-order constellations can have roughly 2000 configurations for a 4-ring 64 APSK constellations. Euclidean distance approach can be used to calculate the optimum configuration, but it is a brute-force method, and calculation process can be very tedious and complex. The approach discussed can find the optimum M-ary constellation for any number of points. It can be used to compare different configurations on account of normalized distance.

Also, when considering the configuration in Rician channel, it is seen when the ring has lesser point and LOS component is more, there is lesser BER.

References

1. Dai L, Wang Z, Yang Z (2012) Next-generation digital television terrestrial broadcasting systems: key technologies and research trends. IEEE Commun Mag 50(6):150–158
2. Thomas CM, Weidner MY, Durrani SH (1974) Digital amplitude phase keying with M-ary alphabets. IEEE Trans Commun 22(2):168–180
3. De Gaudenzi R, Guillén i Fàbregas A, Martinez A (2006) Turbocoded APSK modulations design for satellite broadband communications. Int J Satell Commun Netw 24:261–281
4. Giugno L, Luise M, Lottici V (2004) Adaptive pre- and post-compensation of nonlinear distortions for high-level data modulations. IEEE Trans Wireless Commun 3(5):1490–1495
5. Liolis KP, De Gaudenzi R, Alagha N, Martinez A, i Fàbregas AG (2010) Amplitude phase shift keying constellation design and its applications to satellite digital video broadcasting. In: De Rango F (ed) Digital video. ISBN: 978-953-7619-70-1, InTech. https://doi.org/10.5772/8042
6. Chen JY, Leu CF (2008) A new principle for circular APSK constellation optimization. In: Proceedings of the international conference on mobile technology, applications, and systems 2008
7. Proakis JG (1995) Digital communications, 3rd edn. McGraw-Hill Book Co.
8. Dutta S, Chandra A (2012) Accurate SER expressions for M-ary dual ring star QAM in fading channels. In: Proceedings of international conference on communications, devices and intelligent systems (CODIS), 2012, Kolkata, India, pp 1–4

Big Data Analytics in the Higher Education: Need of the Future

Praveen Mukhia Titimus

Abstract The twenty-first century can be said as the era of engulfing data. From social networking sites to politics, from businesses to education, all sectors of the modern world are flooded with voluminous data. As Alan Kay quotes "this is the century in which you can be proactive about the future; you don't have to be reactive", big data analytics can be the tool and the need of the future to predict the trends, detect the challenges, and leverage the opportunities. Among various sectors like business, social networks, entertainment, politics, educational institutions are seen and expected socially to be more transparent and accountable. With these growing concern and hunger for the academic excellence both to individuals and institutions as well, it becomes inevitable to implement big data analytics in the educational institutions. Its proper implementation can bring the revolutionary development on the education sector. Instead of some inherent challenges, big data analytics can represent customized learning environments to the learners, can reduce potential dropouts, lower the academic risks, complexity could be reduced and can enhance the quality of education system as a whole. The main objective of this paper is to highlight on the emergent need and scope of big data analytics in the educational sector.

Keywords Big data analytics · Educational sector · Proactive future Accountability

1 Introduction

With the advent of information revolution, social media and networks viz; Facebook, Twitter, Whatsapp, Google, 3Gs, 4Gs, etc., are generating huge volume of data every second. Again mobile devices generate data by tracking all objects all

P. M. Titimus (✉)
Department of Computer Science and Application, St. Joseph's College,
North Point, Darjeeling, India
e-mail: titimus.sjc@gmail.com

© Springer Nature Singapore Pte Ltd. 2019
H. K. D. Sarma et al. (eds.), *Advances in Communication,
Cloud, and Big Data*, Lecture Notes in Networks and Systems 31,
https://doi.org/10.1007/978-981-10-8911-4_3

Fig. 1 https://www.
slideshare.net/asertseminar/
big-data-34369979 [5]

the time, like instant messages, voice calling, video conferencing, GPS. Similarly, sensor technologies like, satellites, GPS, radars generate huge data by measuring other sorts of data. This leads to the huge collection of data termed as big data. This leads to systematic management, storage and processing of data which involves tremendous effort and computational as well as storage capability.

The term big data analytics was coined in the year 1990 by John Mashey. Big data analytics involves uncovering hidden patterns, correlations, trends as well as useful information in this huge amount of data. The term "big data" is generally used to describe data sets so large they must be analyzed by computers. Usually, the purpose is to find patterns and connections relating to human behavior and how complex systems function. Analytics generally refers to the process of collecting such data, conducting those analyses, generating corresponding insights, and using that new information to make smarter decisions.

When Tim Berners-Lee shared about data, he wouldn't have imagined growing to an enormous size encircling all spheres of life (Fig. 1).

Big data analytics has a major role to play in all sectors of the modern world. Big data analytics in higher education could be leveraged to enhance creativity, potentiality, problem-solving ability, customized learning environments, reduce potential dropouts, lower the academic risks among students. It can also open up a wide range of possibilities for students and teachers by freeing their minds and setting them to do bigger things rather than fastening to the same pattern of education.

2 Information Technology and Education

Since, information technology is becoming an integral part of life, it plays a vital role in higher education also. An education system involves different layers and processes. Some of the processes may be such as teaching–learning, tools and materials development, curriculum development. The higher education sector is very dynamic and is rapidly changing which leads to the need to adapt tools or techniques rapidly. That is why, it is always under the scrutiny of accrediting agencies and involves continuous monitoring by various stockholders.

Traditional systems are gradually replaced by the devices and gadgets by redefining the learning process. Due to use increasing popularity of use of the devices, many new learning tools like, lectures, e-books, presentation materials are developed with considerable amount of data collectively, which eventually contributed to big data. Many institutions fail to make efficient use of the huge amount of data available [1].

Therefore, big data can be considered as the next frontier in higher education which will revolutionize the same. It would lead to a new wave of technological advances which would help increase academic effectiveness [2].

3 Need for Big Data Analytics

There is an enormous growth of both structured and unstructured volumes of data which can be availed from IT system with the help of big data analytics; otherwise, the risks being swamped by a data overflow. Big data analytics can play a major role in enhancement of the higher education system. On the outset, following issues can be considered in this context:

- **To management resources**: One of the most important needs in today's education institutions is the management of its resources ranging from financial, infrastructural, educational (printed/digital) to human resources.
- **To detect study pattern and individual performance**: In this competitive age, individual performance and excellence is everyone's desire. So, bringing customized study pattern or curriculum is much needed.
- **To increase rate of success of students**: Big data analytics will not only simplify the tasks, but it will help in increasing the success rates of students and institutions as well.
- **To identify the risk at earlier stages**: Big data analytics are needed to detect the risks at the earlier stages.
- **Predict future performance**: As Alan Kay quotes "this is the century in which you can be proactive about the future; you don't have to be reactive", big data analytics can be the tool and the need of the future to predict the trends, detect the challenges, and leverage the opportunities.

4 Challenges for Big Data Analytics

Despite having many opportunities and scope to be the need of the future in educational sector, implementation of big data analytics imposes great challenges. Some of the challenges are as follows.

- **Lack of proper software and technology**: Lack of specific software and technology is impeding its implementation. Big data has become synonymous with Hadoop. Hadoop solves the big data storage problem, but creates another equally troublesome one. As organizations dump more and more data into Hadoop file systems, which lack traditional information architectures, they frequently lose track of what those systems contain. So they cannot find the right data to analyze.
- **Lack of analytic skills**: Many projects fail or suspended indefinitely due to insufficient skills of the analyst. Big data analytics often suffer from unclear objectives and lack of communication which will limit the insights of the projects, if one does not have the right people with right capabilities.
- **Insufficient budget**: Budget is required to kick-start the project of big data analytics, and it will be needed throughout its maintenance and management. Insufficient budget entails great challenges delaying the work, losing the relevance, and limiting the work of the analysis.
- **Finding the correct and related data set**: Inaccessible data do not work at all. Most significantly, one approach of big data analytics is with a well-defined strategy to collect the correct and related data for the success of the organization. The difficulty arises if there is a difference in the data, and then it must be inspected from a different perspective and make the best possible use of relevant insights. One must understand its origin, verify and validate, avoid useless data, discard bad data entirely for the successful analysis.

5 Scope

Big data analytics will create a number of opportunities for the educational institution, administrators, policy makers, educationalists, and also for the learners. These opportunities include [3]:

- **Customized learning environments**: In the past decade, e-learning has evolved from computer-aided instruction, through intelligent system, to smart classrooms, and to mobile learning. Today, e-learning has become learner-centered, personalized learning technologies and pervasive learning. Pervasive learning refers to learning that is available anywhere anytime. It is supported by wireless communication with mobility as well as device independency. Learning wherever and whenever needed is to become possible-learning should not only generate good learning outcomes, but also better learning process from behaviorally, intellectually, and emotionally involved in their learning tasks [4].
- **Decreasing potential dropout rates**: The institutes are still pushing to decrease students' dropout rate using big data analytics. Now with big data, institute can identify students in crisis and intervene before they walk out the doors for the

last time. The big data analytics found that attendance history, class performance, and socioeconomic status were the most accurate predictors of the future dropouts.

- **Web-based learning environments**: As the education sector is started using cutting edge technology, there is need of using the modern teaching–learning tools, which includes—online lectures, e-books, presentation materials which are available across the globe. Along with this, webinars as well as online classes needs to be accessed by the students and teachers.
- **Cost reduction**: The effect of big data on education intelligence is huge. Predictive analytics can be the effective means to minimize the cost of the analysis. Instead of searching every possible known data related with better learning processes, resource management, and policy making, focus can be made only on specific attributes to cut down the test time and overall cost.
- **Faster and better decision making**: Handling big data is not too easy. People need sufficient expertise and tools to detect and analyze the underlying trends. Detection or prediction of proper tools will facilitate the decision-making process in any sector of modern civilization. Integrating big data analysis with traditions systems will generate new insights, thereby improving the efficiency of the system. It will help the following decision-making mechanism of a system:

  (a) **Strategic decisions** are the ones that focus on top-level management that have a potentially wide effect and require a heavier set of recourses for execution.
  (b) **Tactical decisions** are used to evaluate the results of strategic decisions and prioritize them into a plan of action, and those decisions require directed and dedicated resources for execution.
  (c) **Operational decisions** that are intended to execute specific tasks regarding the tactical strategy and prioritized tactics can be integrated directly into operational processes and may only require a small amount of resources to execute.

An example is guiding prospective students to the best courses for their specific desire (Fig. 2).

Fig. 2 https://upload.
wikimedia.org/wikipedia/en/
1/11/Anthony-triangle.JPG
[6]

6   Conclusion

The big data analytics plays a vital role in modern education system. The uses are in various levels and processes of the system right from the teaching–learning materials to decision-making systems. Lots of educational materials and tools are available in World Wide Web (WWW). It can be said that use of big data analytics will reform the complete education system and will give a new direction. Through the proper use of big data analytics, the revolutionary development on the education sector could be achieved. Instead of some natural challenges, big data analytics can represent customized learning environments to the learners, can reduce potential dropouts and failure, and can develop long-term learning plans. All of these are possible through the effective development and use of big data analytics in the educational institutions [3].

References

1. Tulasi B (2013) Significance of Big Data and analytics in higher education. Int J Comput Appl 68(14):0975–8887
2. Big Data (2011) The next frontier for innovation, competition and productivity. James Maniyka, Executive summary, McKinsey Global Institute. http://www.mckinsey.com/mgi/publication/big.data/MGI_big_data_exec_summary.pdf
3. Anirban S (2014) Big Data analytics in the education sector: needs, opportunities and challenges. Int J Res Comput Commun Technol 3(11)
4. Shen L et al (2009) Affective e-learning: using "emotional" data to improve learning in pervasive learning environment. J Educ Technol Soc 12(2)
5. https://www.slideshare.net/asertseminar/big-data-34369979
6. https://upload.wikimedia.org/wikipedia/en/1/11/Anthony-triangle.JPG

Land Capability Classification for Agriculture: GIS and Remote Sensing Approach—A Survey

Shirshak Gurung

Abstract Land capability classification for agriculture assesses the land based on various land capability factors and variables to produce thematic land capability mapping for economical and sustainable agriculture. The main objective of this paper is to survey the important available GIS and remote sensing-based methods proposed for land capability classification. The influence of different capability factors and their sources is also discussed. Furthermore, the scopes of present-day GIS and remote sensing technology for land characterization with important terminologies used are also reviewed.

Keywords GIS · Remote sensing · Soil characteristics · Digital elevation model (DEMs) · Land classification methods · Land use · Land suitability criteria

1 Introduction

Agriculture is one of the most significant activities in sustaining human life on earth. The rate at which land resources are replenished is very slow when compared to the rate at which population is growing [1]. Widespread use of ill-informed and unsuitable cropping practices fuelled by inefficient agricultural farm management techniques have led to the degradation of farmlands, reduced crop production, natural habitat and the ecosystem. All of these contributing factors have facilitated the origin of land capability classification. Land capability classification is the process of formulating and classifying the fitness of land for a defined use [2]. It is about managing changes in the characteristics of crop fields to achieve higher yield using fewer resources thereby having higher profitability and sustained agriculture. Various aspects of the environment such as soil characteristics, climatic factors, vegetation, topography and hydrology influence crop growth and farming success. Detailed information on such parameters provides for accurate analysis on land

S. Gurung (✉)
Department of IT, Sikkim Manipal Institute of Technology, Majitar, Sikkim, India
e-mail: gurungshirshak@gmail.com

© Springer Nature Singapore Pte Ltd. 2019   29
H. K. D. Sarma et al. (eds.), *Advances in Communication,*
Cloud, and Big Data, Lecture Notes in Networks and Systems 31,
https://doi.org/10.1007/978-981-10-8911-4_4

capability and their classification for agriculture. Vital information about land capability help farmers achieves higher sustainable crop yields, higher financial benefits and better environmental conservation [3]. A new farming revolution catalysed by the advent of innovative technologies such as satellites, global positioning systems with high accuracy, smart sensor technologies and a range of IT applications has emerged. Technologies such as high precision satellite remote sensing imagery and Geographic Information System (GIS) have taken land capability classification to new frontiers and greater heights.

There will be an acute food shortage due to the predicted increase in population to 9.7 billion by 2050. The Food and Agriculture Organization (FAO) approximates a need to increase food production by 70% to cope with the rising population. The hunger statistics published by United Nations World Food Programme estimate that about 795 million people of the total 7.3 billion people in the world today were victims of extreme malnourishment as they do not have enough food to live a healthy life. Global environmental problems that include soil acidity and degradation, pollutions, melting glaciers, climatic changes, droughts and floods all affect agricultural activities, thereby making food security the biggest threat to sustaining human life on earth. One of the major problems facing agriculture worldwide is the loss of crop land due to soil erosions, soil degradation, infrastructure construction and other urban amenities in agricultural lands [4]. Thus, the importance of land classification and ultimately land use planning should be seen in the context of cultivable land being a limited and non-renewable resource. Scientifically enhanced land classification methods need to be utilized for the identification of various land uses.

2 Background—GIS and Remote Sensing for Land Capability Classification

A. *GIS*—Geographic Information System makes use of spatial and spectral data from various data sources such as remote sensing datasets to extract quantitative information for agriculture policies and practices in order to provide farmers with the following advantages [5].

- Better crop quality and reduced crop land degradation can be achieved through better agricultural practices and planning with the help of accurate GIS information by avoiding pests, droughts and erosion hazards.
- By cultivating and harvesting different crop type with respect to the information on soil changes of the crop land can maximize yield and also protect the environment.
- Classifying new land parcels for crop cultivation, either to make existing yields more efficient or to increase quantity for our growing populations.
- Identifying and ensuring conservation practices over areas of cultivation that is prone to natural disasters such as drought, flood and soil deficiencies can decrease the risk of financial loss and crop damage.

GIS is a general term used to refer to a large family of independent and inte-grated technologies, processes and methods used to store, manipulate and analyse geographic data. GIS-related applications are used in many fields such as engineering, planning, management, agriculture, insurance, telecommunications and business. A GIS technology utilizes digital information to overcome geographic tasks of spatial and spectral analysis. The success of GIS lies in its ability to integrate variety of data sources like Global Positioning System (GPS) and remote sensing system (RS) through satellite imagery, aerial photographs or ground truths that contain spatial and spectral data. This integrated spatial and spectral information provides decision makers a clear picture of the opportunities and constraints for the defined use [6]. The ability to provide efficient query, analysis and integration makes GIS an ideal scientific tool for land capability classification and thereby enhancing land use planning. Hence, management of agricultural resources based on the information about their potentials and limitations can have a huge impact on sustainable agriculture. Many users of GIS are harnessing the advancements in GIS and satellite technology to create spatial database to arrive at appropriate solutions/strategies for sustainable development of agricultural resources [7].

GIS is playing a pivotal role in enhancing many sectors in agriculture especially in agricultural economic studies where agricultural land use and their changes are one of the areas that agricultural economists are interested in [8]. GIS is a family of software that links digital map with spatial databases. All features on the map (which represent objects in the real world) are related to records in the spatial database, which contains many attribute, value pair to further describe the feature. These features act as a storehouse of information where the digital map stores physical features in the form of points, lines or polygons and the database stores information about them. Thematic land capability mapping for crop production depends on many spatially related factors like climatic variables, vegetation, location specific attributes such as prices of inputs, outputs, household character-istics, land area, parcels of land holdings, water and soil characteristics. The combination of these factors with a set of production technologies that relates inputs and outputs are all provided through GIS [9].

B. *Remote Sensing Products and Free World Datasets*

Remote sensing techniques are being widely used for monitoring of agricultural activities and crop health as they are constrained with special features not common to other economic sectors. They follow strict seasonal patterns with respect to biological lifecycle of crops; it strongly depends on the physical topographic relief (e.g., soil type), as well as climatic drivers and agricultural management practices. All of the features vary in space and time with an additional constraint that most of the agricultural products are perishable [10]. Remote sensing can efficiently help provide a timely and accurate picture of the agricultural activities, as they can provide accurate information over large areas with high revisit frequency.

As compared to (1–8 km) AVHRR data and Landsat data, MODIS data provide for a detailed, large-area LULC classification by providing global coverage with more revisit frequency and intermediate spatial resolution (250 m) [11]. In [12], the author concludes that MODIS time-series data have adequate temporal and radiometric resolution to classify important crop types and crop land use in Kansas, USA. The author in [13] has used the LISS III data to prepare the physiographic soil map for Dehradun, Uttarakhand, whereas the work in [14] has used the IRS LISS III data to effectively map the LULC for the proposed region for land capability classification.

C. DEMs—Elevation/Slope/Topographic Wetness Index

Digital elevation model (DEM) is an electronic model that represents the surface of the earth, and it can be stored and manipulated in a computer digitally [15]. It provides for greater functionalities when compared to the traditional basic qualitative characterization of topography. Various kinds of data relating to processes that are influenced by topography can be derived and manipulated from a DEM. They are used for various kinds of analysis and mapping as they provide quantitative description of landforms and of soil variability. A DEM can be used to extract slopes, aspects, rate of change of slope, drainage network on catchments areas [15, 16]. The information extracted from a DEM can further be analysed with other spatial data and images to improve their capabilities for land capability classification and soil survey mapping [17]. The main procedures involved for manipulation and analysis in GIS are transformation (i.e., from raster to vector data structure), geo-corrections, overlay and interpolation.

3 Land Capability Classification

A. Capability/Suitability Analysis Terminologies

A few of the most commonly used terms are defined to help understand the discussions on land capability classifications [18].

- Capability Factors—These are climatic, soil characteristic, vegetation, hydrological or topographic relief variables which are considered while determining land capability for agriculture. Other capability factors such as the distance to amenities such roads and parks or to hazardous waste sites that may influence land capability may be considered.
- Factor Types—These are a set of possible values or ranges for a particular capability factor. For example, there are five standard depth classes or factor types for use in soil surveys: less than 10 in., 10–20 in., 20–40 in., 40–60 in. and more than 60 in.
- Weights—They are numerical values to quantify the importance of different capability factors when determining land capability for some defined use. For example, the weights for the factor slope maybe 100 as compared to soil

moisture with 50 when determining a particular land use to suggest that slope is twice as much important than soil moisture for that land use.

- Capability ratings—They are numerical values to quantify the importance of different factor types when determining land capability for some defined use. For example, capability rating maybe assigned to each of the above defined soil depth factor types to indicate their relative importance when determining the capability of land for a particular land use.
- Capability Score—They are the overall numerical values to quantify the importance of a location for the defined land use when all of the capability factors have been considered. For example, a location maybe suitable, moderately suitable or not suitable depending on the overall score.

B. *Capability Factors and Their Sources*

The following are the most widely used factors to determine the capability of land for agriculture.

- Slope and Topography—Shape and topographic relief of the region is described by its slope and topography. The measurement of elevation is called topography, and the per cent change in that elevation is called the slope for a certain distance. According to [19], the effect of slope can largely deteriorate soil nutrients due to increasing extent of erosion. The study concluded that slope effects soil, run-off and nutrient losses but crop cover can decrease slope effect and help reduce soil, run-off and nutrient losses to a great degree. Gentle-to-moderate slopes can be used for cultivation with some amount of conservation such as terracing, however steep slopes and ridges are not suitable. Slope and elevation can be extracted from SRTM/ASTER DEMs that are freely available for downloads.
- Soil Texture—This is an important soil property used to characterize soil classes based on their physical texture to help classify crop suitability. They reflect the "feel" of the soil influenced by the soil composition. There are many taxonomy such as USDA soil taxonomy, WRB soil classification system and the UK-ADAS system. They are characterized by the composition of sand, silt, clay and loam present in the soil. It influences many soil physical properties such as water holding capacity and inherent fertility.
- Soil Depth—Amongst the different soil characteristics, soil depth plays an important part in determining crop roots growth and the amount of water and air in the soil. Soil depths which are not deep and have lithic contact may restrict root growth, thus creating a suboptimal environment of shallow volume soil which negatively impacts the growth and yield of the crop. A suboptimal measurement of effective soil depth will frequently alert a scientist to possible production problems. The amount of clay present in the soil greatly influences soil depth. The soil composition depends on the nature of the parent material on which soils were developed [14]. The remotely sensed products offered freely in the Internet by International Soil Reference and Information Centre (ISRIC) as soil grid data are rich sources for soil depth, organic content and different soil types.

- Soil Wetness—Soil water or wetness in the amount of water held within the soil pores. Soil wetness or soil moisture availability depends on the quantity of rain, frequency of rain, quality of soil texture to hold moisture and the soil depth [20]. The concept of moisture availability index [MIA] was developed for this purpose. In [20], the author proposed that MIA be the standard for measuring water deficiencies or excesses. He suggested the following classification of MIA:

0.00–0.33   very deficit
0.34–0.67   moderately deficit
0.68–1.00   somewhat deficit
1.01–1.33   adequate
1.34 and above   excessive

The Digital Elevation model (DEM) can be used to extract topographic information that influences many terrain-related processes. One such process that can be calculated is the topographic wetness index, which estimates the soil wetness based on the topography [21].

- Organic Content—Organic matter in the soil refers to the partly decayed plant and animal residue. It is important to have a good amount of organic matter in the soil composition as they provide for an ideal source of plant nutrient. Upon decomposition of organic matter plant, nutrients are released in an available form. The adsorption sites for crops nutrients are represented by stable organic fraction, called humus. Humus is also important in maintaining soil structure, soil tilth and reducing soil erosion. Soil organic matter varies spatially with natural soil variability and soil management. Amount of organic matter content indicates the soil health and its suitability for agriculture. Very often in land capability classification average organic carbon content of the soil types is taken to represent the organic matter content in each soil type and capability weights are assigned accordingly [14]. Soils with low organic matter have poor soil structure and hold little water.
- Soil Erosion—Soil erosion is one of the most important factors involved in destroying many fertile agricultural soils around the world. Hence, predicting the erosion factor for the crop land can be of great help in evaluating land capability.
- Ancillary Factors—These are the additional factors that strongly influence the land capability index for agriculture such as road connectivity, human settlements, pollutions and presence of toxic materials.

C. *Land Classification Methods*

Land capability and suitability classification are the basis of proper land use. Since the introduction of the first classification criteria in the 1930s by the Soil Conservation Service of the USA, several such criteria and methods have been developed. Although capability and suitability are relatively interchangeable terms, but as per the guidelines by FAO, capability is used in a more general term to quantify the fitness with considerations to prevent land degradation whereas the

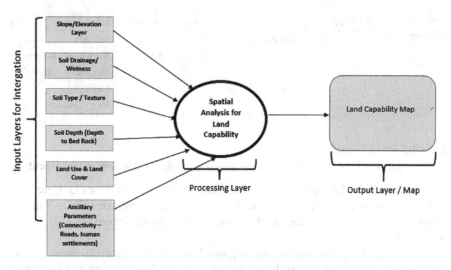

Fig. 1 Basic process model for land capability classification (agriculture)

latter is used to consider the fitness of a given type of land for a more specific or defined use. Land suitability can be classified for present condition of the land suitability factors (actual land suitability) or after improvement of certain land suitability factors (potential land suitability) [22]. The capability of land to sustain agriculture is analysed through the interaction of different factors in GIS as shown in Fig. 1 with the help of spatial analysis techniques such as advanced spatial analysis tool, Python as command line or script and Map algebra through Python or raster calculator.

Numerous systems of land classification have been proposed for use in different regions with varying conditions, the prominent being that proposed by [23, 24]. The FAO has also proposed a land capability/suitability classification framework that has four different categories: Orders, Classes, Subclasses and Units. There are two orders (S and N) which reflect the kind of suitability (S for suitable and N for unsuitable land). The framework allows for total freedom in determining the number of classes within each order. However, the use of only three classes within order Suitable (S) and two classes within order Not Suitable (N) is the optimum recommendation [25]. The different classes are indicated by numerical values in sequence of decreasing suitability within the order. Therefore, classes are the degree of suitability within the orders as shown in Table 1.

The subclasses are indicative of the limitations or improvement required within classes. They are symbolized as follows:

c: Climatic conditions.
t: Topographic limitations.
w: Wetness limitations.
n: Salinity (and/or alkalinity) limitations.

Table 1 Land suitability order and class (FAO 1976)

Order	Class	Symbol	Label
Suitable (S)	1	S1	Suitable
	2	S2	Moderately suitable
	3	S3	Minimally suitable
Not suitable (N)	1	N1	Unsuitable
	2	N2	Very unsuitable

f: Soil fertility limitations.
s: Physical soil limitations (influencing soil/water relationship and management).

For example, a land unit maybe represented by the symbol S3w (2). Here "S" represents Order (Suitable), the value 3 after the letter S represents Class 3 (minimally suitable), "w" indicates the Subclass w (wetness limitation), and (2) represents unit 2.

Tideman in [26] suggests two broad groups for land classification, namely land suitable for cultivation that includes classes 1–5, and land not suitable for cultivation but very well suited for forestry, grassland and wildlife. The land suitable for cultivation can be determined through common parameters such as soil texture, depth, slope and erosion. Further local limitation such as salinity, alkalinity, water logging, climate may also be considered. In [27], the author proposes a qualitative evaluation of land to determine land suitability for rice and wheat cultivation based on four variables, like Nitrogen–Phosphorous–Potassium (NPK) status, soil reaction (pH), organic carbon (OC) and soil texture. The proposed study is an integrated approach using remote sensing and GIS techniques.

The author in [28] establishes the role of remote sensing (RS) and GIS in spatial planning for agriculture in North West Province of South Africa. The primary data were obtained from soil samples, and secondary information was acquired from the remotely sensed SPOT 5 data. The capability factors taken were soil texture, soil depth, clay fraction, pH and land use/land cover. The final classification map was produced using weighted overlay method in ARCGIS software package with all the factor layers for sunflower and sorghum. The author uses Eq. (1) for the assessment of land capability.

$$
\begin{aligned}
\text{LSP} = {}& 0.2(\text{LU})_{i=1-4} + 0.1(\text{SL})_{i=1-4} + 0.2(D)_{i=1-3} + 0.2(T)_{i=1-5} \\
& + 0.2(\text{pH})_{i=1-312} + 0.05(C)_{i=1-3} + 0.05(\text{DR})_{i=1-3}
\end{aligned} \tag{1}
$$

where LSP is the numerical index of the land suitability, LU is the land use/land cover factor (with factor types 1–4), SL is the slope factor (with factor types 1–4), D indicates depth factor (with factor types 1–3), pH is the soil pH (with factor type 1–3), T represents texture (with factor types 1–5), C indicates clay percentage (with factor types 1–3), and DR represents drainage (with factor types 1–3). The author obtained the final capability map by assigning capability ratings to the different factor types. For example, slope for factor type = 0–4% (weight = 0.1, rating = 10, land capability = good) and for slope for factor type = 4–8% (weight = 0.1, rating = 8, land capability = good), etc.

In [13], the author provides land capability classification and crop suitability for wheat and mango in Dehradun, Uttarakhand. The work utilizes soil texture, soil drainage, coarse fragments, erosion hazards and slope for the study which classifies land into eight classes grouped into two suitability groups/orders, i.e. land suitable for cultivation and land for other uses. However, certain critical observations are made such that some of the physiographic units classified as suitable are overlapping with settlements and river beds when visually compared with the given classified LULC maps. Basic socio-economic factors such as human settlements and road connectivity that may have negative or positive impact on agriculture needs to be appropriately weighed in when modelling the land capability classification for agriculture. The author in [14] proposes a method for land capability classification using five parameters that include slope, LULC, soil texture, organic content and soil depth. Weighted Aggregation method in ARC/INFO GIS is used to analyse the different parameters that have been assigned suitable knowledge-based weights as in Eq. (2) to determine the Land Suitability Potential for agriculture as shown:

$$LSP = 0.2(LU)_{i=1-12} + 0.2(ST)_{j=1-5} + 0.1(SL)_{k=1-6}$$
$$+ 0.25(OC)_{l=1-5} + 0.25(D)_{m=1-4} \tag{2}$$

where LSP is the numerical index of the land suitability, LU is the land use/land cover factor (with factor types 1–12), ST is the soil types (with factor types 1–5), SL is the slope factor (with factor types 1–6), OC represents the organic carbon content (with factor types 1–5), and D indicates the depth factor (with factor types 1–4). The superscripts i, j, k, l and m indicate subclasses based on importance in determining the land suitability.

4 Conclusion

Integration and analysis of spatially related land capability factors provides for financially profitable and sustainable agriculture. With the growing population growth-rate pressure and the ever-increasing global environmental problems, there is an urgent need to have easy access to scientifically accurate and timely data about land capability and their management. This need further becomes a reality with the advancement in remote sensing and GIS technologies which have been reviewed in this paper.

References

1. Davidson DA (1992) The evaluation of land resources. Longman Scientific and Technical, Harlow
2. Marsh SP, MacAulay TG (2002) Land reform and the development of commercial agriculture in Vietnam. Aust Argribusiness Rev 10
3. www.cema-agri.org/page/precisionfarming
4. www.wfp.org/hunger/stats
5. www.enviromentalscience.org/agriculture-scienc-gis
6. Ghafari A, Cook HF, Lee HC (2000) Integrating climate, soil and crop information: a land suitability study using GIS. In: Proceedings of 4th international conference on integrating GIS and environmental modelling, Alberta, Canada
7. Venkataratnam L (2002) Remote sensing and GIS in agriculture resource management. In: First national conference on agro-informatics
8. Shrestha KB (2003) Nepal, agricultural diversification and international competitiveness. APO
9. Baniya N (2008) Land suitability evaluation using GIS for vegetable crops in Kathmandu Valley/Nepal. Ph.D. thesis submitted to Humboldt University of Berlin
10. Food and Agriculture Organization of the United Nations (FAO) (2011) Global strategy to improve agricultural and rural statistics. Report No. 56719-GB, FAO, Rome, Italy
11. Justice CO, Townshend JRG, Vermote EF, Masuoka E, Wolfe RE, Saleous N, Roy DP, Morisette JT (2002) An overview of MODIS land data processing and product status. Remote Sens Environ 83:3–15
12. Wardlow BD, Egbert SL, Kastens JH (2007) Analysis of time-series MODIS 250 m vegetation index data for crop classification in the US central great plains. Remote Sens Environ 108:290–310
13. Bhandari S, Jhadav ST, Kumar S (2013) Land capability classification and crop suitability assessment in a watershed using RS and GIS—a case study of watershed in Dehradun, Uttarakhand. In: 14th Esri India user conference
14. Bandyopadhyay S, Jaiswal RK, Hegde VS, Jayaraman V (2009) Assessment of and suitability potentials for agriculture using a remote sensing and GIS based approach. Int J Remote Sens 30(4):879–895
15. Brough PA (1986) Principle of geographical information systems for land resources assessment. Oxford University Press, 194p
16. Brabyn L (1997) Classification of macro landforms using GIS. ITC J 97(1):26–40
17. Lee KS, Lee GB, Tyler EJ (1988) Thematic mapper and digital elevation modelling of soil characteristics in hilly terrain. Soil Sci Soc Am J 52:1104–1107
18. Pettit CJ, Klosterman RE, Delaney P et al (2015) The online what if? Planning support system: a land suitability application in Western Australia. Appl Spat Anal Policy 8:93. https://doi.org/10.1007/s12061-015-9113-7
19. Khan F, Naeem WM, Bhatti AU (2001) Effect of soil steepness and wheat crop on soil, runoff and nutrient losses in eroded land of Malakand Agency, NWFP, Pakistan. Sarhad J Agric 23:101–106
20. Hargreaves GH (1975) Moisture availability and crop production. Published in American society of Agricultural and Biological Engineers, St. Joseph, Michigan
21. Mandel A, Ferrero VO, Graser A, Bruy A (2016) QGIS 2 cookbook. Packt Publishing Ltd., UK, pp 190–195. ISBN:978-1-78398-496-1
22. Ritung S, Agus F, Hidayat H (2007) Land suitability evaluation with a case map of Aceh Barat District. Indonesian Soil Research institute and World Agroforestry Centre, Bogor, Indonesia
23. Storie RE (1954) Land classification as used in California. In: Fifth international soil conference, Leopoldville

24. Ricquier J, Bramao DL, Cornet JP (1970) A new system of soil appraisal in terms of actual and potential productivity. Soil Resource Development and Conservation Service, A.O., Rome
25. Mishra A (2007) Land suitability classification for different crops. Orissa Rev
26. Tideman EM (1996) Watershed management, guidelines for Indian conditions. Omega Scientific Publications, pp 38–41. ISBN:81 85399 34 4
27. Halder JC (2013) Land suitability assessment for crop cultivation by using remote sensing and GIS. J Geogr Geol
28. Kabanda T (2015) Land capability evaluation for crop production using remote sensing. GIS and Geo statistics in Rietfontein North West Province of South Africa. Geo UERJ, Rio de Janeiro 26:2–21

Review on Vehicular Radar for Road Safety

Additi Mrinal Singh, Soumyasree Bera and Rabindranath Bera

Abstract The paper is a survey of vehicular radar for road safety. Vehicular radar is also called automotive radar. It is introduced for avoiding collision and for reducing human and economic losses in road accidents. Automotive radar functions to detect objects and obstacles in its surrounding area and to prevent collision. Its other applications include speed sensing, predictive crash sensing, collision mitigation, and automatic braking which are addressed in this survey. This paper presents recent approaches done in automotive radar and trend and requirements of autonomous driving.

Keywords Automotive cruise control · Collision mitigation · Long-range radar
Short-range radar · Laser radar · Autonomous driving

1 Introduction

The automotive radar for road safety has been under development for several decades, and even from early beginning, its aim was collision avoidance. The first car radars using an active frequency doubling reflector on the backside of target cars having frequency of 10 GHz were constructed in early 1960s in the USA. But shortly, it was realized that it is not sufficient for heavy accidents. So it was required to reduce the size of the antenna, and thus, the frequency was increased from 10 to 35 GHz and was road-tested [1–3]. In early 1970s, radar developments were started

A. M. Singh (✉) · S. Bera · R. Bera
Department of Electronics and Communication Engineering,
Sikkim Manipal Institute of Technology, Majitar, Sikkim, India
e-mail: adittimrinal18@gmail.com

S. Bera
e-mail: soumyasree.bera@gmail.com

R. Bera
e-mail: rbera50@gmail.com

© Springer Nature Singapore Pte Ltd. 2019
H. K. D. Sarma et al. (eds.), *Advances in Communication,
Cloud, and Big Data*, Lecture Notes in Networks and Systems 31,
https://doi.org/10.1007/978-981-10-8911-4_5

at microwave frequencies (66–71 GHz). Since today, automotive radar system tests are being done on 77–79 GHz frequency.

In 1999, Mercedes-Benz became the first car manufacturer to introduce radar-based autonomous cruise control (ACC) system in the S class. Besides ACC, 77 GHz long-range radar sensors are also used. 77 GHz provides for safety system like predictive crash sensing, and collision mitigation in many mid- and high-range vehicles [2–6].

2   Automatic Cruise Control System

In 1995, ACC was commercialized for the first time in Japan. Japan initiated ACC system with LIDAR sensor, and European and US companies initiated radar-based ACC. Mercedes introduced the 77 GHz "Distronic" into the S class in 1999. Other vehicles in which an ACC system was introduced are Jaguar (XKR, XK6), Cadillac (STS, XLR), BMW 7 series, Audi A8, and VW Phaeton. ACC is also available in Audi A6, BMW 5 and 6 series, Honda (Accord, Inspire, Odyssey), Lexus (LS, GS), Mercedes E, CL, CLK, SL class, Nissan (Cima, Primera), and Toyota (Harrier, Celsior) [4]. Radar sensor is used in adaptive cruise control system to detect the objects in its track and surrounding. An ACC system maintains the vehicle speed set by the driver, and it changes the speed to preceding vehicle's speed, to maintain safe distance if the vehicle in front is moving at slower speed. Then, again, it automatically switches to set speed when traffic clears. There has been past research for ACC systems aimed to design a vehicle following distance control system using linear approximation and linear control logic [4–8].

3   Automotive Sensors

To track objects and to monitor the movements of the vehicle in front as well as the movement of the surrounding objects to predict and mitigate collisions, different sensors are used. Major problem of automotive cars is the high cost of the sensor. Sensors present in automotive vehicles are:

(i) **Radar Sensor**: Radar is the most commonly used sensor in automotive detection applications for detecting and tracking. It uses electromagnetic waves to find the range, direction, Doppler velocity, altitude, and angle of objects. A single antenna is used to emit and receive the radar waves, and the sensor constantly switches between the sending and receiving mode. Radar cannot determine the details or color of the target but it works in all weather condition, so it is weather independent [9].

(ii) **LIDAR Sensor**: LIDAR is laser radar which is similar to millimeter wave radar in its basic function. Lidar provides range, Doppler velocity, and angle measurements. Lidar is used for obstacle detection, obstacle avoidance, and to navigate safely through environments. To image object, Lidar uses ultraviolet, visible, or near-infrared light [10].

(iii) **Cameras**: Cameras are used in automotive system to observe lane marking, to observe people and vehicles in the lane. Different cameras used for these are:

    (a) **Single Camera Systems**: or mono-cameras are used to observe the lane markings in lane-keeping aid systems.

    (b) **Stereo Camera System**: It is a high fidelity camera which gives 3D imagining of the object. It is able to use the difference in the images within one camera shot to detect every type of obstacle, from loads that have fallen onto the road to people and animals, and can find their size and the distance between them.

    (c) **Infrared Cameras**: In automotive systems, IR cameras are introduced for night vision systems as driving becomes more difficult after sunset. The IR sensors have a unique ability of measuring the temperature of objects, and it is sensitive to wavelength similar to the normal body temperature of humans and animals. IR cameras could be so useful in classifying objects and providing accurate angle measurements. In automotive collision mitigation (CM) systems, IR cameras are not used, but they are much useful to the driver for night vision [11].

4 Automotive Radar Applications and Frequency Selection

Radar is used in autonomous vehicles due to its ability to work in rainy and foggy conditions which other sensors like laser and sonar cannot do. Thus radar is weather independent and measures velocity and range both. Radar also has advantage that works effectively even being mounting behind a plastic bumper [4].

4.1 76–77 GHz LRR System

The introduction of 77 GHz system was started in Germany in the early 1980s as a worldwide standard for long-range automotive radar (LRR). The aspired goals for the LRR sensors are to cover a higher distance range from less than 1 m to up to 300 m and to have opening angle up to ±30° and a relative velocity range of up to ±260 km/h. It is based on narrowband operation with narrow beamwidth. In 1998, the Mercedes-Benz DISTRONIC system came in operation at this frequency. Later, DISTRONIC PLUS was introduced which has combination of two 24 GHz short-range radar sensors (SRR) and one 77 GHz long-range radar sensor (LRR). In a critical situation or when the traffic situation requires heavier braking, then DISTRONIC PLUS is able to perform, and the driver is warned by audio and signals, which reduces driver's burden. But the responsibility for safely handling the vehicle remains with the driver under all circumstances [4, 12–14].

4.2 24 GHz SRR System

The 24 GHz frequency band was adopted for short-range automotive radar sensors (SRR). Its main aspect is range accuracy and is based on wideband operation, which gives a wide view of the surroundings of the car at moderate speeds. And as it does not measure the angle of the detected objects, single antenna element is sufficient for this. For short-range applications, UWB sensors are preferred in SRR because of their low-cost perspectives and high resolution. Compared to 77 GHz, 24 GHz does not a have higher frequency and it has low resolution and less Doppler sensitivity. The FCC allocated automotive UWB short-range radar systems with the band of 22–29 GHz having a maximum mean power density of −41.3 dBm/MHz [15–17]. SRR sensors operate in continuous wave mode (CW, FMCW, and FSK), in pulsed mode (pulse, pulse Doppler), or in coded radar with spread spectrum techniques (pulsed, CW, pseudo-noise). SRR sensors can enable a variety of applications like:

- Stop and go functionality
- Collision warning
- Blind spot monitoring
- Parking aid (forward and reverse)
- Lane change assistant
- Rear crash collision warning

4.3 76–81 GHz Short-Range Radar

Higher-frequency radar systems tend to perform better because they are more reliable and more accurate. For short-range radar applications, the resolution should be high, so a wide bandwidth is required. Therefore, the use of short-range vehicular radar operating in the frequency band of 76–81 GHz is maintained by the European Commission (an ETSI Standard) in 2004. And it was concluded by EU in 2013 that 24 GHz range will not be used further. The main advantages of the 79 GHz frequency band are that a single technology can be used for all applications, radar devices can be much smaller, and the risk of mutual interference problem will be low as it requires low emission power [16–18].

There are four main reasons for changing 24 to 76–81 GHz:

1. Shifting to the 76–81 GHz range allows developing radar modules that will be used for all automotive radar types from LRR via MRR to SRR. Hence, it will be a cost-effective change.
2. Future vehicle safety system demands higher resolution and accuracy in space and time.
3. Vehicle integration and sensor packaging demand minimization, while enhancing sensor performances.
4. With the increase in technology, mutual interference has become an issue. A large bandwidth having frequency hopping or other frequency separation methods has to be mandatory for interference mitigation and is worldwide available only between 76 and 81 GHz.

Besides these requirements, 77 GHz LRR and 24 GHz SRR are already in use and 76–81 GHz SRR is in process. European Union Commission is encouraging trends of employing even higher-frequency bands than 77 GHz for automotive use like 158 GHz for ACC and 122 GHz for near-range parking sensor.

5 Future Trends

ACC was developed for collision avoidance only. With the increase in technology, requirement has increased. Today, ACC system is already in trend for autonomous driving. Countries like Europe, China, and Japan are in track of autonomous driving. In Japan, the organizer of the autonomous driving project is the MLIT (Ministry of Land, Infrastructure, Transport, and Tourism). The European slogan of the running decade in direction of automotive sensor development is:

2011 -2020 - the decade of action for road safety

The vision of vehicle motion and safety of Germany and Europe is to have accident-free driving by the year 2020 and fully autonomous driving by 2030 and onward [11–21] (Fig. 1).

Fig. 1 Visions for vehicle motion and safety [19]

6　Requirements for Autonomous Driving

(1) **Static and Dynamic Environment Sensing**: For driving as well as for parking functions, localization and navigation are mandatory. Today, available sensors are not sufficient to distinguish between static, dynamic, and slow-moving objects like animals, human beings. It is a major issue for autonomous driving and to reduce accident rate which may be solved using Doppler processing [22, 23].

(2) **Decision-Making**: For autonomous driving, ACC system is required to have good decision-making capability. As mention above, after sensing its environment, the system must be able to take decision distinguishing between objects. If two or more vehicles are approaching at 70 km/h, then it should have decision-making capability to slow down or apply brake or to change lane to avoid collision. Along with signal understanding, the system also has good decision-making capability [22].

(3) **Ego-motion Estimation**: In vehicles, for velocity and yaw rate (rotational speed around the height axis), sensors are usually available. But sometimes due to large rotational velocities around the other axes due to braking or bad roads, tilt or roll motion is not measured. So the estimation of the ego vehicle's motion is a key requirement for automated vehicles localization.

(4) **Mutual Interference Problem**: If there are two radars operating with the same modulation scheme and same frequency band, mutual interference will occur. The beat signal of another radar transmitted signal with the actual radar transmitted signal will be interpreted as a target echo which leads to false alarm and unsafe decision. This can be mitigated by signal processing method like PN-coded FMCW or spreadspectrum technique [24].

(5) **Cost of Radar and Sensors**: High cost of automotive radar is a major issue.

(6) **Weather**: Increase in frequency leads to weather problem. Heavy rain or snowfall can interfere with cameras and laser sensor.

7　Conclusion

Automotive radar is being used since many decades. Many technologies have been introduced, and most are in use practically while some are only theoretically successful. The system developed is not perfect specially for autonomous driving. Improvement is required to be done in image processing for collision avoidance and blind spot detection. Mutual interference in radar is also an issue. To mitigate accident rate, slow-moving object detection and distinguishing between static and dynamic object are also required. Automotive radar is being used today for autonomous driving and is coming as future trend so better system is required to develop.

References

1. Groll HP, Detlefsen J (1996) History of automotive anti-collision radars and final experimental results of a MM-wave car radar developed on the technical university of munich. Technische Universitat Munchen, Microwave Department, Munchen, Germany, pp 13–17
2. Schneider M, Bosch R (2005) Automotive radar—status and trends. In: German Microwave Conference. IEEE, pp 21–24
3. Mu L et al (2009) Research on key technologies for collision avoidance automotive radar. In: Intelligent Vehicles Symposium. IEEE, pp 223–236
4. Wenger J, DaimlerChrysler AG (2005) Automotive radar—status and perspectives. In: CSIC Digest, Germany. IEEE, pp 21–28
5. Miyta S et al (2010) Improvement of adaptive cruise control performance. EURASIP J Adv Signal Process 1–8
6. Meinel HH (2014) Evolving automotive radar—from the very beginnings into the future. In: The 8th European conference on antennas and propagation. IEEE, pp 3107–3114
7. Kumar R, Pathak R (2012) Adaptive cruise control—towards a safer driving experience. Int J Sci Eng Res 3(8):1–6
8. Baizhuo W (2001) Survey of automotive radar technologies. Nanyang Technological University, Singapore, pp 1–5
9. Skolnik MI. Radar handbook, 2nd edn. McGraw-Hill Inc., pp 1–4
10. Mertz C et al (2013) Moving object detection with laser scanners. J Field Robot 17–43
11. Srivastava S et al (2015) Collision avoidance system for vehicle safety. Int J Sci Res Dev 3 (04):1730–1732
12. Luo T-N, Wu C-H, Chen Y-J (2013) A 77-GHz CMOS FMCW frequency synthesizer with reconfigurable chirps. IEEE Trans Microw Theory Tech 61(7):2641–2647
13. Hasch J, Topak E, Schnabel R, Zwick T, Weigel R, Waldschmidt C (2012) Millimeter-wave technology for automotive radar sensors in the 77 GHz frequency band. IEEE Trans Microw Theory Tech 60(3):845–860
14. Adachi K (2006) Proposal of a target headway distance method and applying this method to a car for adaptive cruise control system. Proc Jpn Soc Mech Eng 06:23–28
15. Jeong SH et al (2012) Technology analysis and low-cost design of automotive radar for adaptive cruise control system. Int J Automot Technol 13(7):1133–1140
16. Meinel H, Dickmann J (2013) Automotive radar: from its origins to future directions. Microw J 56(9):24–40
17. Dickmann J, Klappstein J et al (2015) Present research activities and future requirements on automotive radar from a car manufacturer's point of view. In: MTT-S international conference on microwaves for intelligent mobility. IEEE, pp 1–4
18. https://itunews.itu.int/en/3935-Future-trends-for-automotive-radars-Towards-the-79GHz-band. note.aspx
19. Bhushan V, Goswami S (2016) An efficient automotive collision avoidance system for Indian traffic conditions. Int J Res Eng Technol 5:114–120
20. Fölster F, Rohling H (2005) An automotive radar network based on 77 GHz FMCW sensors. Hamburg University of Technology, Germany. IEEE, pp 1–6
21. Evans R, Farrell P, Felic G, Duong HT, Le HV, Li J, Li M, Moran W, Skafidas E (2014) Consumer radar—opportunities and challenges. In: 11th European radar conference (EuRAD), Oct 2014, pp 5–8
22. Dickmann J et al (2016) Automotive radar the key technology for autonomous driving: from detection and ranging to environmental understanding. In: IEEE radar conference, pp 1–6
23. Hahn M, Dickmann J (2014) Autonomous maneuvering with radars. IWPC-Workshop, Detroit, USA
24. Luo TN, Wu CH, Chen YJ (2013) A 77-GHz CMOS automotive radar transceiver with anti-interference function. IEEE Trans Circuits Syst I Regul Pap 60(12):3247–3255

Landmark-Based Robot Navigation: A Paradigm Shift from Onboard Processing to Cloud

Prerna Rai and Dhruba Ningombam

Abstract Cloud robotics is one of the important and upcoming technologies of the twenty-first century. It is in actual fact a computing system which is distributed in nature, and the processing is performed over the cloud, for computation of huge amount of tasks. Cloud services help robots to find its application, both in the indoor or outdoor environment for various purposes in our everyday life. Before the advent of cloud, the robots were mostly independent and performed computation and storage in the onboard computer or microchip. This led to huge overhead on the processing power of the robot. Therefore, the survey is organized around two major aspects of robot navigation, i.e. robotics with onboard information and robotics with cloud infrastructure. This paper highlights the change in the technologies for robot navigation and the benefits of cloud infrastructure.

Keywords Cloud robotics · Landmark · Navigation · Robot

1 Introduction

Robot is a machine designed to perform one or more tasks autonomously using its intelligence. Their essential elements are sensing, movement, energy, and intelligence. The history of robots dates back to the year 1920 when Karel Capek first introduced Rossum's Universal robot [1] in his play. In the year 1941, for the first time, the term robotics was used by Isaac Asimov referring to it as a science and technology of robot. Asimov in his book "I, Robot" went on to formulate the Three Laws of Robotics, set of rules which have now been widely adopted in the field. According to [1], robotics can be defined as the branch of various engineering fields

P. Rai (✉) · D. Ningombam
Department of Computer Science and Technology, Sikkim Manipal Institute
of Technology, Majhitar, Rangpo, East Sikkim, India
e-mail: pthulung@gmail.com

D. Ningombam
e-mail: dningombam@gmail.com

© Springer Nature Singapore Pte Ltd. 2019
H. K. D. Sarma et al. (eds.), *Advances in Communication,
Cloud, and Big Data*, Lecture Notes in Networks and Systems 31,
https://doi.org/10.1007/978-981-10-8911-4_6

such as mechanical engineering, electrical engineering and computer science. They deal with the design, production, functioning and application of robots. The major components of robots are [2]:

(a) Mechanical structure: wheeled platform, arm, motor, DC motors, AC motors, stepping motors, servomotors.
(b) Sensors: light sensors, touch sensors, sound sensors and acceleration sensor.
(c) Control systems: power supply, control system such as logic circuit and microcontroller which is characterized by speed, size and memory.

Microcontrollers are the core elements of many robots. With the rapid advancement in the design and capacity of microprocessor and other techniques, the capabilities of robots have increased to greater extent. Today robotics is in a new venture for an application which demands advanced intelligence. Robotic technology is introduced with a wide variety of technologies such as machine vision, force sensing which has vividly changed the speed and efficiency of the new system [3].

It can be concluded from various researches that robotic application started its journey from automated industries and moved towards the augmentation of robots into intelligence. However, the technology used in traditional robotics has been limited by the innate physical constraints. They are especially not meant for large-scale environment due to onboard computations of task as they have limited computing capabilities. Therefore, with the introduction of the World Wide Web, the complexity in computing task by robots decreased dramatically. The performance and efficiency of robot has seen an exponential increase with the introduction of cloud and its services.

Hence, the survey is based on the change in the paradigm beginning from the independent robot to the cloud-based architecture for achieving navigation of mobile robots based on landmark, taking advantage of the intensive computation, huge storage and availability of other shared resources of the cloud [4].

Cloud infrastructures provide high consistency, larger storage capacity, and stable power system, reuse of hardware, high scalability and better resource utilization. These properties make cloud a better approach towards the functioning of robot. In particular, the dynamic scalability property of cloud is exceedingly useful in robotics [5]. Thus, the term cloud was combined with robotics in the year 2010 by James Kuffner and called cloud robotics [6].

For the efficient functioning of robot, there are four major benefits from the cloud to the robot [7]. They are as follows:

(1) Big data: access to remote data store of imagery, maps and other data.
(2) Cloud computing: access to on-demand parallel computing for statistical analysis, learning and motion plan.
(3) Collective robot learning: robots sharing and exchanging knowledge with other robots having different hardware and software.

(4) Human computation: It is the use of crowdsourcing and call centres. It provides on-demand access to remote human guidance for analysing image, learning and controlling and error recovery.

The study presented here is based on the navigation of robot using different landmarks. It shows the different techniques in navigating the robot based on landmarks since navigation is one of the fundamental problems that exist in cloud robotics and is particularly of two types: local and global navigation [8]. Local navigation deals with navigation on the scale of few metres, while global navigation deals with navigation on a larger scale where robot cannot view the goal state. In this type, the main problem is obstacle avoidance. In this paper, the literature survey task is performed to find a difference in techniques used by different researchers and the limitations of the same techniques. From the study, it is found that most researchers rely either on navigation of robot using a predefined map or generating a map on the go. Some of the approaches rely on techniques of localization and others on robot navigation based on landmark [9].

2 Literature Survey

During the last few years, several researchers have performed studies that have focused broadly on two different aspects of robotics for navigation of robot. One of the aspects is the use of onboard computation and storage, and the other is using cloud infrastructure or cloud services. The study focuses on the advancement in the technology that changed the world of robot navigation and how there has been a gradual shift from independent robots to cloud robots.

2.1 Robotics with Onboard Information for Navigation

Several researchers have worked on the navigation of the robot using various techniques based on landmark. The very initial technique used for navigating in an environment is using landmark-based maps where robot has camera and sensors to identify landmarks and measure the angle subtended by the landmark as stated in [9]. The position of landmark in the external coordinate system is given on the map, and linear positioning estimation algorithm is used to localize robot while in motion given a noisy data input. For the purpose of landmark identification, the system compares the pixel values of pictures obtained from the images captured by the camera to the picture that is stored in the database. The pixels provide information on the bearing of one landmark relative to another landmark. However, the major challenge was the presence of large outliers in the images obtained from the camera. This affects the positioning and orientation of robot and causing misidentification of landmarks in a large environment [1]. Furthermore, the course of concurrently tracing the location of a mobile robot to its surrounding and generating the environmental map has been an essential research topic in the field of robotics. Therefore, with the introduction of simultaneously localisation and map building (SLAM), the possibility of identifying landmarks and navigation of mobile robot in larger and unmodified environment has substantially improved. In the paper [10], vision-based SLAM algorithm has been proposed. The system uses scale-invariant feature transform (SIFT) and was developed by Lowe in the year 1999 for object identification The image features do not vary with image transformation, scale, rotation and also to some extent invariant to illumination changes. Due to these properties, it makes them appropriate landmark for robust SLAM. Moreover, SIFT features are considered to be good natural visual landmarks for tracking objects over an extensive period of time from a different view and angle. The system compares the SIFT landmark with the stored SIFT characteristics for the landmark identification required for robot navigation using the transformation technique and builds a 3D map. However, an error exists with image features, and analysing error is important. Solution to the problem is determined using variant approaches such as the ones based on Kalman filtering [11]. This method is also known as linear quadratic estimation (LQE). LQE is an algorithm that uses a sequence of measurements observed over time and results in more precise estimates of unknown variables compared to those based on a single measurement alone. Though it provides precision, it cannot identify the features distinctively. Therefore, in later years another approach was developed based on 'Markov localization' [12]. It is a probabilistic algorithm that provides accuracy for representing probabilistic distribution of different kinds. But the major drawback of this technique is that it requires considerable computation power for updates therefore is not suitable for large environment. Monte Carlo localization (MCL) is one of the suitable ways to the fix the above-stated constraint. MCL, also known as particle filter localization, is an algorithm developed in computational robotics to localize using a particle filter. The algorithm finds the location and direction of robot as it moves in an environment with the map provided. It also senses the environment. Apart from the technique mentioned, there were many on the line but none of the techniques deal with change in topology. Therefore, in the paper [8], an algorithm has been proposed where the robot is allowed to navigate precisely and reliably using a sensor network in dynamic environment. The navigation occurs node to node, starting from the start node to the goal node as shown in Fig. 1.

The principle under which the network and robot interact is based on the communication served by network. The network also provides sensing and computation medium for the robots. For a robot to navigate, the navigation goal is provided and the node that is nearest to the goal triggers navigation field computation. Thereafter, each node probabilistically determines best possible direction for navigation purpose. The proposed system uses adaptive delta per cent algorithm for determining the node that is nearest to the robot.

Thus, the system enhances the navigation of robots using many networked sensors but the major challenge lies in pre-programming every node with

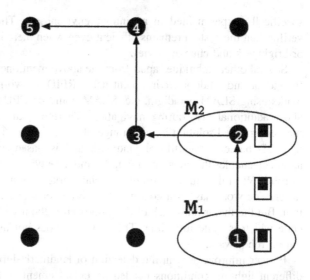

Fig. 1 Node-wise navigation based on sensor network

information about its neighbours, and moreover, each node in the network should probabilistically know its neighbour.

For any robot to navigate, the basic requirement for a system would be to have a complete navigation of mobile robot that integrates both local and global navigation systems. For example, Vector Field Histogram* algorithm (VFH*) in [13] combines the VFH+ and A^* search for both local and global path planner. Similarly, many other local obstacle avoidance algorithm such as dynamic window (DW) and nearness diagram (ND) [14] are combined to perform navigation for complete system, while A^* algorithm is combined with DW algorithm to develop D^* algorithm. However, several shortcomings were identified. The mobile robot was trapped in a loop when it reaches a dead end. Currently D^* search algorithm is the most widely and feasible for use in dynamic environment. The use of D^* algorithm can obtain information about global environment from any position. Therefore, the paper [15] introduces a complete navigation method called DVFF. This algorithm combines D^* algorithm, with virtual force field (VFF) approach which is an obstacle avoidance algorithm. It also makes use of ultrasonic sensors without assuming prior knowledge about the environment. The paper, however, does not take into consideration, the information about topology of the obstacle's position.

In mobile robot navigation based on landmark, vision is one of the major elements of interest. Visual landmarks have played a major role in making efficient navigation compared to the output of odometers. The work presented in [16] aims at "natural" visual landmarks built over SLAM paradigm. The extraction of landmark is focused mostly on quadrangular objects, e.g., ventilators, windows, posters, cupboards. These selected landmarks are the basic man-made structures. The paper highlights two main directions regarding visual landmark-based navigation system which is described for topographical navigation, and to handle ambiguous landmark, the Markovian localization is implemented. The system proposed is specifically experimented in an indoor environment. The experiment conducted verifies that the system remains efficient even when there is a variation in lightning or brightness and change in view.

Several other techniques apart from the above mentioned have been investigated for indoor and outdoor environment using RFID, networked sensors, visual landmarks using SLAM paradigm, etc. SLAM could be CPU intensive and would not allow additional task during navigation. Therefore, the introduction of artificial marker provided solution in wide range of environment. The paper [17] makes use of a 2D barcode as artificial landmark and is based on computer vision. The approach enables the creation of map for indoor environment and allows navigation of robot. With the introduction of artificial marker, the navigation is comparatively faster. The work has been conducted for both corridor and multiple room navigation. But the major drawback of this approach is the use of visible black and white code which is quite impeding. The approach does not use movement orientation during robot walk.

Further improvement in the detection of landmark during robot movement and different lighting conditions needed an enhancement in the previous work. This enhancement has been proposed in Zhang et al. [18]. The system is based on 2D code landmark for the process of localization of robot in an indoor environment. In the situation where the information of the robot's surrounding is not provided, the researcher proposes a blind search and vision-based search strategy for navigation of robot. Vision-based search strategy considerably reduces the amount of time consumed on locating a landmark compared to blind search strategy. However, the system does not consider obstacle avoidance.

Despite the fact that the technology in the field of robotics has evolved rapidly, most of the existing robots are not fully automatic and unreliable [19, 20] for surveillance purpose. The possibility of distinguishing between a path and an obstacle and making a robot autonomous using Kinect sensor has been proposed in the paper [17]. The paper proposes low-cost autonomous mobile robot for safe navigation by developing a software implementation comprising of both obstacle avoidance algorithm and navigation algorithm. The system uses filtering and clustering process to the 3D point cloud converted from the acquired video, but the main problem that arises is in the quality of image captured and also the use of vision-based sensor that may incur certain cost. Apart from all the research, the related work presented in [17] focuses on the development of an indoor autonomous mobile robot that can be used as a tour guide. Many other tour guide robots were developed. Most researchers used RFID technology while some made use of laptops and smartphones. The proposed system deals with navigation, sensor integration and control. Combining a smartphone application with robot technology using microcontroller is the main achievement of this robot. In the proposed system, the robot is autonomous and is programmed to recognize QR code using ultrasonic range sensors. The system is based on Bluetooth technology in order to send information provided in the QR code to the tour guide robot. For the purpose of robot navigation at higher speed, the QR reader had difficulty recognizing the QR code. To meet the challenge, the system devised a mechanism that detects a circle-shaped marker to slow down the speed for reading the code properly. But the research work could be extended further for more robust turning control of the robot.

2.2 Robotics with Cloud Infrastructure for Navigation

During the last few years, several researchers show how landmarks can be used for mobile robot navigation. The use of 2D code or QR code and other artificial markers paved a way ahead towards the use of growing technology. As per the survey done in [21], the field of robotics has advanced considerably since 1995. There are over 5 million service robots like the Roomba vacuuming house and services, over 3000 robots assisting surgeons in operating rooms, etc. With the introduction of Kinect 3D camera by Microsoft, came a major breakthrough for robotics. By the year 2012, the major outburst in the field of robotics came with a huge initiative taken by US National Robotics.

The robots application is extended further to solve the problem of uncertainty. As of today, robots are still not intelligent enough to work under unknown environments. Hence, the cloud is considered to be the key to the new generation of robots. Cloud computing for robotics is considered to be useful in many ways which include saving of power and energy of robot, increase storage capacity, efficient resource utilization. Cloud is a service provider to robots especially when the amount of data to be stored increases vigorously due to growing objects around us and also due to uncertainty and huge computation. Therefore, the solution to the problem lies in cloud robotics. Till this moment, robots are viewed as self-contained system with limited computation and memory. Today, cloud robotics makes use of Internet where robot can access and exchange code and data via wireless networking.

By the year 2013, the field of cloud robotics became one of the truly required fields of research. The research performed in [22] deals with cloud robotic paradigm for enhanced navigation of autonomous vehicle. The system proposes data fusion technique with advanced sensing systems for automated guided vehicles (AGVs). It is generally used for automated transportation of raw materials in industrial environments. The AGV makes use of sensing system that enables reliable obstacle detection gathering information about environment, but the major challenge is in classifying objects, and the other challenge is to navigate freely by increasing the flexibility. Therefore, to meet with these challenges, the system proposes a centralized system known as Global Live View cloud service which contains updated information about all the entities and thereafter a real-time global map is produced.

The era of cloud robotics began with "Remote brained robots" [23], a work by Inaba et al. in the year 1997. By the year 2009, World Wide Web for robots was envisioned which led to the announcement of a RoboEarth [24] project. RoboEarth allows the robots to share and communicate data and also provide opportunity to learn from each other. This steady growth of robotics towards cloud has helped in enhancement of robot navigation in the most efficient manner.

In the year 2015, another related work is presented by Salmeron [5]. The proposed system shows the possibility of using a vision-based navigation assistant on an external cloud. It also shows the improved performance of robot navigation when compared with the onboard processing of the robot. The system proposes a robot with a stereo camera and sends frames to the teleoperator and to the process that extracts 3D point cloud (3DPC). The 3DPC is then sent to the navigation assistant node. The node also receives input command from local command interface and are empirically translated to the desired linear and angular speed (V_j, ω_j). Once the information about obstacle and robotic velocity is obtained correctly, $V_{\max}$ is computed to avoid obstacle collision. From the analysis made for computation offloading in this work [5], it is found that 3DPC extraction and navigation assistant in the cloud provides high scalability and efficient computation offloading compared to the processing in the robot without cloud.

In the same year, another related work based on RoboEarth project was undertaken. The paper [25] describes the vision of RoboEarth project in designing knowledge-based system. RoboEarth is a knowledge base for robots that converts a simple robot into an intelligent one. In this paper, the researchers state that a robot functioning in an environment for the first time can get the information previously stored by any other robot using semantic robot description language (SRDL). The project also provides collision-free navigation capability using A^* algorithm and ORM algorithm. Hence, with RoboEarth, the working of simple robots with typical computation and communication facilities can be furthered.

In the aforementioned works, several researchers have used landmarks which are used to provide local space information. They have provided the robot with cloud. Furthermore, even if the current system bases its study about robots in realistic environments, the paper [26] includes referential markers to provide information to robot real-time environment. The paper aims to provide a dynamic way to configure mobile robots and allows navigation in complex and wide indoor environments that are not known in advance. The system is based on three main modules. They are environmental tags, a Cloud SaaS and a Maps Supervisor. The environmental tags such as AR Tag and QR Tag is used by the system. These tags provide information about the remote resource to the robot. The Cloud SaaS is the storehouse of all the information about the environment, and finally, the Maps Supervisor takes the output from the tag detection module for the assessment of the map. Depth cameras are used for tag detection mechanism and are extremely cheap and widely used in robotics, but the system needs to further work on the use of scene recognition techniques in order to identify features instead of QR codes. Also, this method modifies prestored information using multiple robots.

3 Analysis of Robotics Without Cloud and with Cloud

Table 1 as depicted below provides an insight towards the change in the efficiency and performance of the robot with respect to the amount of task performed. It further illustrates an improved tolerance to fault because of the decentralized architecture, and above all, it elucidates how the cost-effective nature of the cloud has made the transition of trends towards a new era.

4 Techniques Used

Table 2 shows the change in the use of methods and techniques used for the robot navigation using landmarks suitable for different environment. It highlights how major changes came by the year 2009, with the introduction of RoboEarth. Other projects like DaVinci [27] also play a vital role for the success of cloud to be used in the field of robotics. By the year 2013, technology had already stepped into a new era, an era of cloud integrated with robotics known as cloud robotics. The year 2010 is marked as the era of cloud robotics. Several techniques were introduced thereafter, and today by 2016, researchers are still on a verge of making robots more dynamic and applicable in real environment.

Table 1 Analysis between robotics with and without cloud

S. No.	Parameters	Robotics without cloud	Robotics with cloud
1	Architecture	Centralized	Decentralized
2	Scalability	Scalable only with networked robot	Highly scalable
3	Heterogeneity	Homogeneous in nature	Supports heterogeneity
4	Security issue	Secured as the data are onboard the robot	Measure should be taken while using cloud
5	Communication bandwidth	Not applicable	Speed of communication heavily relies on communication bandwidth
6	No of task performed	Limited	Can be extended using cloud service
7	Cost	Expensive	Cost-effective
8	Fault tolerance	Low tolerance	High tolerance

Table 2 Techniques used till date

S. No.	Year	Navigation technique used	Suitable environment
1	94–97	Linear positioning estimation algorithm Uses pixel to pixel comparison	Small environment
2	1999	Markov localization	Dynamic environment but impractical for large environment
3	1999	MonteCarlo localization	Large environment
4	2002	Vision-based SLAM algorithm using SIFT	Global environment
5	2004	Sensor network	Dynamic and small environment
6	2005	DVFF approach	Dynamic environment
7	2007	Use of natural visual landmarks with SLAM paradigm	Indoor and local environment
8	2009	Introduction of RoboEarth project	Networked robot
9	2013	• 2D barcode • Global live view cloud service	• Indoor • Dynamic
10	2014	Use of 2Dcode and vision-based search strategy	Dynamic environment
11	2014	Use of Kinect sensor	Indoor environment
12	2015	• 3DPoint cloud extraction of image and cloud-based robot navigation assistant • Use of open-source RoboEarth and communication using SRDL	• Static environment • Enhances communication both in small and large environment
13	2016	Use of referential marker QR and AR tags. Uses cloud services	Real and dynamic environment

5 Conclusion

This paper provides a chronological study of the major changes that took place in the use of methods and techniques for robot navigation using landmark. The study also delivers findings that authenticates the use of cloud with robotics and subsequently augments the performance and efficiency of robots by shifting the head of robot to cloud, or by using the cloud storage. Along with its benefit, there are limitations of cloud as well. Sometimes, there are difficulties in controlling robots motion as it depends on sensors, feedback of controllers and Internet connections. Many a times, cloud-based applications can get slow due to high latency responses or network problem. If robot is totally dependent on cloud, a small fault in network can leave the robot brainless. It can be hacked easily as real-time execution is required for the same.

But in future, robots can be made to store the interactions made with cloud so that if the connection goes offline, the robots can retrieve the stored data from their own memory in order to sense and react to their surroundings.

References

1. Betke M, Gurvits L (1997) Mobile robot localization using landmarks. IEEE Trans Robot Autom 13(2):251–263
2. Kurfess TR (2004) Robotics and automation handbook. CRC Press, pp 1–3. ISBN 9780849318047
3. How robot works science.howstuffworks.com. Access date 20.09.16
4. Robotics technology trends www.automation.com/library/articles-whitepaper. Access date 21.09.16
5. Salmeron-Garcı J (2015) A tradeoff analysis of a cloud-based robot navigation assistant using stereo image processing. IEEE Trans Autom Sci Eng 12(2):444–454
6. Zhu X, Qiu C, Tao Y, Jin Q. Cloud based localization for mobile robot in outdoors. NSFC No. 60905052
7. Kehoe B, Abbeel P. A survey of research on cloud robotics and automation. IEEE Trans Autom Sci Eng
8. Batalin MA, Sukhatme GS, Hattig M (2004) Mobile robot navigation using a sensor network. In: Proceedings of 2004 IEEE international conference on robotics and automation, ICRA'04, vol 1. IEEE
9. Topological mapping for mobile robots using a combination of sonar and vision sensing. In: Proceedings of the AAAI, 1994, pp 979–984
10. Se S, Lowe D, Little J (2002) Mobile robot localization and mapping with uncertainty using scale-invariant visual landmarks. Int J Robot Res 21(8):735–758
11. Kleeman L (2016) Understanding and applying Kalman filtering. Electrical and Computer Systems Engineering Monash University, Clayton, Access date 20.09.16, pp 11–16
12. Fox D, Burgard W, Thrun S (1999) Markov localization for mobile robots in dynamic environments. J Artif Intell Res 11:391–427
13. Ulrich I, Borenstein J (2000) VFH*: local obstacle avoidance with look-ahead verification. In: 2000 IEEE international conference on robotics and automation, San Francisco, CA, 24–28 Apr 2000, pp 2505–2511
14. Fraichard T (2007) A short paper about motion safety. In: Proceedings 2007 IEEE international conference on robotics and automation. IEEE
15. Djekoune AO, Karim A, Toumi R (2009) A sensor based navigation algorithm for a mobile robot using the DVFF approach. Int J Adv Robot Syst 6(2):97–108
16. Hayet J-B, Lerasle F, Devy M (2007) A visual landmark framework for mobile robot navigation. Image Vis Comput 25(8):1341–1351
17. Lee SJ et al (2014) Autonomous tour guide robot by using ultrasonic range sensors and QR code recognition in indoor environment. In: IEEE international conference on electro/information technology. IEEE
18. Zhang S et al (2014) Initial location calibration of home service robot based on 2-dimensional codes landmarks. In: Control conference (CCC), 2014 33rd Chinese. IEEE
19. Correa DSO et al (2012) Mobile robots navigation in indoor environments using Kinect sensor. In: 2012 Second Brazilian conference on critical embedded systems (CBSEC). IEEE
20. Benavidez P, Jamshidi M (2011) Mobile robot navigation and target tracking system. In: 2011 6th international conference on system of systems engineering (SoSE). IEEE
21. Guizzo E (2011) Robots with their heads in the clouds. IEEE Spectrum, pp 16–20

22. Cardarelli E et al (2015) Cloud robotics paradigm for enhanced navigation of autonomous vehicles in real world industrial applications. In: 2015 IEEE/RSJ international conference on intelligent robots and systems (IROS). IEEE

23. Inaba M (1997) Remote-brained robots. In: Proceedings of international of joint conference on artificial intelligence, pp 1593–1606

24. Kehoe B et al (2015) A survey of research on cloud robotics and automation. IEEE Trans Autom Sci Eng 12(2):398–409

25. Riazuelo L et al (2015) Roboearth semantic mapping: a cloud enabled knowledge-based approach. IEEE Trans Autom Sci Eng 12(2):432–443

26. Limosani R (2016) Enabling global robot navigation based on a cloud robotics approach. Int J Soc Robot 1–10

27. Arumugam R, Enti VR (2010) DAvinCi: a cloud computing framework for service robots. In: 2010 IEEE international conference on robotics and automation, 3–8 May 2010

From Cognitive Psychology to Image Segmentation: A Change of Perspective

Anju Mishra, Priya Ranjan, Sanjay Kumar and Amit Ujlayan

Abstract Image segmentation is a complex and essential task used in many computer vision applications. The problem of image segmentation can essentially be formulated as a grouping problem which in its simplest form tries to group the pixels of image into distinguished regions of interest so that further processing of the extracted regions can be achieved. This work proposes an image segmentation model which is inspired by the findings in cognitive psychology theories to divide the image into separate coherent regions. The proposed work tries to correlate between human and machine cognition by studying the segmentation process under the light of psychology of human vision.

Keywords Image segmentation · Foreground extraction · Cognitive psychology
Machine vision

1 Introduction

Image segmentation is the task of decomposing an image into its constituent regions of interest, i.e., objects and the background present in that image. A lot of work [1–6] has already been done in this direction, and various types of approaches have been formulated and proposed by researchers across the globe to cope up with this

A. Mishra (✉) · P. Ranjan
Amity University Uttar Pradesh, Noida, Uttar Pradesh, India
e-mail: amishra1@amity.edu

P. Ranjan
e-mail: pranjan@amity.edu

S. Kumar
Oxford Brookes University, Oxford, UK
e-mail: skumar@brookes.ac.uk

A. Ujlayan
School of Vocational Studies and Applied Sciences, Gautam Buddha University,
Greater Noida, India
e-mail: iit.amit@gmail.com

© Springer Nature Singapore Pte Ltd. 2019
H. K. D. Sarma et al. (eds.), *Advances in Communication,
Cloud, and Big Data*, Lecture Notes in Networks and Systems 31,
https://doi.org/10.1007/978-981-10-8911-4_7

61

simple task. For many applications, separating figure and ground is a necessary and inherent step for further processing, e.g., developing intelligent machines and cameras that can automatically separate the object of interest based upon the context of application. For a human, the task of separating figure from its ground is very simple and inherent, from computational or machine point of view, the same task is considered to be a difficult one, and as the nature or the characteristics of the environment (e.g., noise and illumination variations and variations in camera angles and perspective) change, the task is transformed into a complex problem. From a psychological perspective, the problem can be defined as a grouping problem in which the visual system tries to group the visual stimuli based on different properties of these stimuli, e.g., intensity, contrast, distance, similarity of color, texture.

The Gestalt school of Psychology [7–11] follows that perceptual grouping plays a vital role in human visual perception, and based on these thoughts, many researchers have formulated various algorithms which follow closely the line of perceptual grouping. Since perceptual grouping can be linked most naturally to the problem of figure–ground separation (i.e., segmentation process), all these proposed algorithms tried to group image pixels based on different pixel properties—be it the color, texture, or contrast between adjacent pixels. An efficient image segmentation algorithm must have two important properties:

(a) It must capture perceptually important regions within an image.
(b) It must be computationally faster in terms of running time so as to be applicable in real-time scenario.

2 Related Work

The latest computer vision algorithms follow one or more of the following mathematical models [2, 4, 6, 12–15]:

Graph-based Approaches: Models the set of image pixels as a graph and segmentation process to be a graph partitioning problem.

Clustering Methods: Models image pixels to belong to different clusters or groups and segmentation process to be the problem of finding non-overlapping coherent groups of pixels.

Appearance-Based Models: Models the image as having different appearances for figure and ground and segmentation process to be the problem of finding key appearance features, e.g., key feature points.

Classification Techniques: Attach a class label with each image pixel to find different areas/regions/patterns inside the image.

Thresholding and Histogram Methods: Focus on forming the bins of pixels based on intensity value and try to find a threshold value suitable to divide the image into segments.

Edge Detection Methods: Segment the image into edges and background.

Region-Based Methods: Partition the image into separate regions.

Template Matching: In this method, segmentation is achieved by creating templates (representation of objects) and then matching them against the input image to find the presence of object in question.

Many psychologists have proposed different viewpoints supported by experiments to prove that visual perception process is either a top-down process or bottom-up process or a combination of these processes [16]. According to the Gestalt theory, visual perception is a global phenomenon which starts from local characteristics of visual stimuli. This leads to the perception of those structures which are constituted from elementary perceptual stimuli but are more stable and strong in terms of global properties. The winning visual structure is constituted as 'figure' and the loser as 'ground.' The Gestaltist has proposed the laws of visual perception which are classified into two categories:

(i) *Local or elementary grouping laws and*
(ii) *Global grouping laws.*

The process of visual perception starts by considering local characteristics such as similarity (of color, shape, texture, etc.), constant width, vicinity [7–9]. Whenever the points in a neighborhood have common characteristics, they get grouped and form visually extended objects. Since all the elementary grouping laws work simultaneously, many overlapping visually extended (global) perceivable structures are formed simultaneously. Finally, global grouping laws governing these extended (global) structures compete with each other and the winning law which governs the most stable structure contributes to the perception of that structure as an object.

The Gestalt view of perception [17] suggests a bottom-up approach for figure–ground perception, where elementary grouping laws work first and form higher-level groups, and in the next step of refinement process, these higher-level groups are synthesized by application and collaboration of global gestalt laws; as a result, different interpretations for figure and ground in the same image are possible.

Psychologists have made constant efforts in this direction and proved experimentally that visual cognition and perception are not only a bottom-up process but sometimes it is a top-down process or a mixture of both. Apparently, prior information and cues or previously learned facts play an important role in the visual cognition process and the domain knowledge and past experiences also speed up the whole process of visually perceiving objects from a scene [10, 11, 16].

3 Perceptual Analysis of Existing Computer-Based Segmentation Methods

By making use of these psychological findings, computer-based algorithms can be formulated to perform image segmentation. [7, 9] has identified that the gap between human and computer vision is due to the fact that human vision primarily depends on the physiology as well as the psychology of brain system while all computer-based methods and systems primarily focus on the physiological aspects of human brain and try to imitate the functioning of the brain system. The most commonly available computer-based segmentation techniques can be broadly classified into three different categories inspired by psychological findings (Table 1):

(i) *Top-down approaches,*
(ii) *Bottom-up approaches, and*
(iii) *Knowledge-based or hybrid approaches.*

The working principle of popular automated image segmentation techniques is explained below which clearly shows that all these techniques got their intuition from perceptual grouping principles proposed by Gestalt which says that the process of visual perception starts by considering local characteristics such as similarity (of color, shape, texture etc.), constant width, vicinity. Whenever the points in a neighborhood have common characteristics, they get grouped and form visually extended objects.

1. *Graphs*: Graph-based approaches to image segmentation try to create a graph of image pixels wherein pixels represent the vertices of graph and these are linked by edges between them based on some similarity criteria.
2. *Clustering*: These methods group the pixels in different clusters of interest based on pixel properties and the neighborhood around them.
3. *Appearance-Based models*: In these methods, the appearance of object region is created (e.g., by extracting key feature points of object region) to model the object shapes.
4. *Classification*: The methods that use a classifier fall into this category. The classifier learns to divide image pixels into coherent segments by utilizing supervised or unsupervised learning methods.

Table 1 Classification of computer-based segmentation methods according to psychological findings

Top-down approach	Bottom-up approach	Learning-based/knowledge based/ hybrid approach
Appearance models	Graph-based	Template matching
Classification	Clustering	Interactive techniques
Region splitting	Thresholding	
	Edge-based	
	Region growing	

5. **Threshold**: In thresholding technique, one or more intensity values (generally called 'the threshold' value(s)) are identified and all the intensity values greater or equal to threshold value are grouped together (generally by setting them, e.g., 1) and all others are set to a different intensities, say 0. In this manner, all the image pixels are divided into two or more groups which correspond to object(s) and ground.

6. **Edge**: The edge-based methods use the contrast information and find the edges and boundaries of object regions. Normally, closed boundaries indicate the location of objects in image and rest of the information is treated as background.

7. **Region**: Region-based approaches start by choosing initial seed points from image space and then growing or accumulating more similar and neighborhood points into the same region.

8. **Template Matching**: In this method, initial cues of actual objects are created and saved and these are called the 'templates,' and then by using some sort of matching function, the similarity between the object in question and the template is decided.

4 Proposed Model

The study of visual psychology enabled researchers to get deeper insights into human cognitive process. It has been concluded that visual cognition is not only about seeing or sensing the environment from visual organ but the perceiving of the objects is a complex process which involves previous knowledge and cues to make the overall process faster and effective. Based on these facts, a generalized automated image segmentation model must have the following components (Fig. 1):

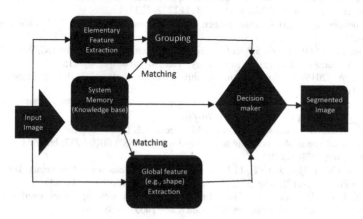

Fig. 1 Proposed model for image segmentation

(a) *An elementary feature extraction and grouping module,*
(b) *A global feature extraction module,*
(c) *A system memory module to hold past knowledge, and*
(d) *A decision maker module to select the most stable candidate segmentation.*

5   Conclusion and Future Scope

The proposed work is an attempt to link psychological findings with computer vision. The proposed model makes use of past knowledge to select between the possible interpretations of object which may help in achieving higher segmentation accuracy. The future work will be to incorporate this model for developing an automated image segmentation system.

References

1. Richtsfeld A, Mörwald T, Prankl J, Zillich M, Vincze M (2014) Learning of perceptual grouping for object segmentation on RGB-D data. J Vis Commun Image Represent 25(1):64–73, ISSN 1047-3203. http://dx.doi.org/10.1016/j.jvcir.2013.04.006
2. Cheng MM, Mitra NJ, Huang X, Torr PHS, Hu SM (2015) Global contrast based salient region detection. IEEE Trans Pattern Anal Mach Intell 37(3):569–582
3. Renninger LW, Malik J (20014) When is scene identification just texture recognition? Vis Res 44(19):2301–2311. ISSN 0042-6989
4. Felzenszwalb PF, Huttenlocher DP (2004) Efficient graph-based image segmentation. Int J Comput Vis 59:167. https://doi.org/10.1023/B:VISI.0000022288.19776.77
5. Liu T, Yuan Z, Sun J, Wang J, Zheng N, Tang X, Shum HY (2011) Learning to detect a salient object. IEEE Trans Pattern Anal Mach Intell 33(2):353–367
6. Boykov Y, Veksler O, Zabih R (2001) Fast approximate energy minimization via graph cuts. IEEE Trans Pattern Anal Mach Intell 23(11):1222–1239
7. Wertheimer M (1923) Untersuchungen zur Lehre von der Gestalt. II, Psychol Res 4(1):301–350
8. Wertheimer M (1958) Principles of perceptual organization. In: Beardslee DC, Wertheimer M (eds) A source book of gestalt psychology, Van Nostrand, Inc., pp 115–135
9. Parkin A (2016) Explorations in cognitive neuropsychology. Taylor & Francis. ISBN 9781317715795
10. Advances in Computer Vision, Volume 1 by C. Brown Psychology Press, 2014, Taylor & Francis, ISBN 1317767667, 9781317767664
11. Driver J, Davis G, Russell C, Turatto M, Freeman E (2001) Segmentation, attention and phenomenal visual objects. Cognition 80(1–2):61–95, ISSN 0010-0277. http://dx.doi.org/10.1016/S0010-0277(00)00151-7
12. Borji A, Cheng MM, Jiang H, Li J (2015) Salient object detection: a benchmark. IEEE Trans Image Process 24(12):5706–5722. https://doi.org/10.1109/tip.2015.2487833
13. Gu X, Deng JD, Purvis MK (2016) Image segmentation with superpixel-based covariance descriptors in low-rank representation. CoRR abs/1605.05466

14. Dhanachandra N, Manglem K, Chanu YJ (2015) Image segmentation using K-means clustering algorithm and subtractive clustering algorithm. Procedia Comput Sci 54:764–771, ISSN 1877-0509. http://dx.doi.org/10.1016/j.procs.2015.06.090
15. Zhang H, Fritts JE, Goldman SA (2008) Image segmentation evaluation: a survey of unsupervised methods. Comput Vis Image Underst 110(2):260–280
16. Vecera SP, Farah MJ (1997) Is visual image segmentation a bottom-up or an interactive process? Percept Psychophysics 59:1280–1296 [PDG]
17. Koffka K (1935) Principles of gestalt psychology, international library of psychology, philosophy, and scientific method, vol 20. Harcourt, Brace and World

An Exploration in Perception-Based Digital Media Processing: A Psychological Perspective

Shanu Sharma, Priya Ranjan and Amit Ujlayan

Abstract The computer vision field deals with the problem of understanding the scene or features in images of real world with the help of image processing and pattern recognition techniques. The main complication in this task is that the objects present in the images may have different appearances to the camera due to illumination effects, camera position, shadows, types of camera, etc. Nevertheless, with the advancement of technologies, today computer vision has provided reliable methods for various tasks like object classification, action recognition, autonomous driving, scene analysis, highlights extraction in videos and many more. But the problem of automatic qualifying is that how well people perform these actions has been largely unexplored. Human visual system and cognition can outperform the performance of computer vision algorithms. The objective of this paper is to highlight the state of the art of various psychological views of human visual perception in computer vision methods that have been found to operate well and that led up to the above-mentioned capabilities.

Keywords Visual perception · Psychology · Video processing
Image processing · Computer vision

S. Sharma (✉)
Department of CSE, ASET, Amity University Uttar Pradesh, Noida, Uttar Pradesh, India
e-mail: shanu.sharma16@gmail.com

P. Ranjan
Department of Electrical and Electronics Engineering, ASET,
Amity University Uttar Pradesh, Noida, Uttar Pradesh, India
e-mail: pranjan@amity.edu

A. Ujlayan
School of Vocational Studies and Applied Sciences, Gautam Buddha University,
Greater Noida, India
e-mail: amitujlayan@gbu.ac.in

1 Introduction

In the last couple of decades, a digital multimedia revolution has been experienced where TV and cinemas have gone digital, and laptops, tablets, smartphones, etc., are having high-quality media streaming over the Internet. Digital videos and images bring broadcasting, cinemas, computers and communication industries together in a truly revolutionary manner. A single device can serve as a personal computer, a high-definition TV and a video phone at the same time. This advancement of digital multimedia has generated a lot of digital contents like movies, news, TV shows, sport videos that can be easily recorded and distributed to anyone and anywhere using satellite perception and can be used by us in our daily life in the form of high-definition TV, in malls, in classrooms, in mobile phones, etc. We can now capture a live video on a mobile device, apply digital processing on it and can transfer it to any place. Other applications of digital media processing can include medical imaging, surveillance of military, law enforcement and intelligent highway systems. Due to this vast amount of video generation and its applications, there is a high demand of good video processing techniques which can be applied to various types of videos and can improve the existing video processing-based applications.

1.1 Computer Vision-Based Video Processing

Video processing is a field of computer vision for automatic interpretation of digital videos using computer-based algorithms. Although humans are generally very fast in interpreting the digital videos, the development of computer algorithms for the same task is highly evasive and so is an active research area in computer vision. The rapid growth in the field of multimedia information and video technologies generates a lot of research interests among researchers for the visual feature-based video retrieval, analysis and other video processing technologies. The basic problems in video processing include

- Searching for events in videos,
- Object extraction like extraction of all people in a scene,
- Removing vibration and jitter in a video,
- Spatially or temporally aligning video captured from multiple cameras.

Due to explosive amount of videos, the traditional approaches of video processing are not able to cope up with it, so new approaches of video processing are need to be developed (Table 1).

Table 1 Various applications of automated video processing

Video indexing and retrieval	Video annotation	Video tracking
Video stabilization	Real-time analysis of scene conditions in videos	Reconstructing 3D scene information in video
Tracking of objects in video	Personalized interactive online video	Video summarization
Video analysis for surveillance	Video-quality evaluation	Video compression
Video processing in compressed domain	Multi-camera video fusion and processing	Objective and subjective video-quality estimation
3D multi-view video compression	Video segmentation	Video enhancement

1.2 Human Perception and Its Characteristics

Perception is the process of brain to organize, identify and interpret information received form sensors for the representation and understanding of environment. It is the ability of a person to see, to hear or to get aware of something through different senses. Perception cannot be seen as a passive acceptance of signals as it can be modified by learning and expectations. The process of sensing and perception can be explained with the help of Fig. 1.

Environmental stimulus is anything that we can see, hear or touch or smell, a part of environment stimulus, focus attention on this stimulus which then excites the receptors, e.g. the visual stimulus forms the image on the retina. Transduction is

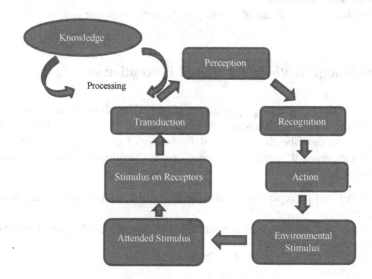

Fig. 1 Human perceptual process

the process of transforming one energy into another, and like in case of visual perception, the image on the retina generates electrical signals in tens and thousand receptors in eyes.

Attended stimulus—is a part of environmental stimulus, focus attention on this stimulus.

Stimulus on receptors—the attended stimuli excite the receptors. Neural processing is the process of transferring the electrical energy to brain by interconnected neurons. This electrical energy transforms in brain to conscious sensory experience called as perception. Recognition and action are the important outcome of perceptual process.

2 Human Visual Perception and Its Characteristics

Visual Perception: Vision is the perception of 3D world from its 2D partial projections onto the left and right retinas of eyes. Vision is the only sense that provides accurate spatial information from a distance. Vision is a residual perception. What you see is what you get. Visual perception can be understood by

- From psychological point of view
- From physiological point of view
- Use of both approaches to create a complete and consistent view...

Physiological Approach
Molecules → Neurons → Circuits and brain areas → Brain

Psychophysical Approach
Individual features → Group of features → Objects → Scene

3 Psychological Views for Visual Perception

Perception is the process of getting aware of something through our senses or the way in which something is understood or interpreted. The main problem for psychologist is to describe the process of formation of perceptual experience after receiving the physical energy from sense organs. Main theoretical issues among psychologist are:

- Some say that perception depends directly on the information present in stimulus only.
- Some say that perception is not direct but it depends on the perceiver's belief and previous knowledge.

Based on these two assumptions, perception can be understood using

- Top-down processing and
- Bottom-up processing.

3.1 Top-Down and Bottom-Up Processing View of Perception

Bottom-up processes are driven by sensory information from the physical world. In case of visual perception, it is carried out from retina to visual cortex. Bottom-up processing defines low-level information which can be used to build up high- or top-level information, e.g. identifying the low-level feature shape for performing high-level process of object recognition.

Top-down processes actively use and extract information through senses from the physical world and are driven by our knowledge, expectations and goals, etc. Top-down processing can be defined as the recognition of pattern through the use of contextual information, e.g. it is easier to understand the writer's objective by reading the whole paragraph rather than reading the terms separately.

This controversy that perception is a bottom-up or top-down process, was discussed by two famous theories of psychologist, Gibson's (1966) direct bottom-up theory and Gregory's (1970) constructionist indirect top-down theory.

A. James Gibson (1966) Direct Theory:

According to James Gibson's theory, perception is direct and can be understood using bottom-up process only. This psychologist has explained perception with following characteristics.

- Perception is directly proportional to sensation means what you see is what you get.
- No high-level processing is required as the information received through sensors about shape, size and distance is sufficient to percept the environment.
- Analysis of sensory information can be done in one direction only. So perception is a bottom-up process only.

Three important components of Gibson's theory to explain visual perception are:

- Optic flow patterns,
- Invariant features,
- Affordances.

Optic flow patterns: The optic array and changes in it contain important information about the motion of an object. It explains that if flow appears to be coming out from a particular point—means perceiver is moving towards that point and if flow is moving towards that point means perceiver is moving away from that point.

Invariant features: It explains that the pattern or structure of gradients provides sufficient information about the environment, as it can be seen that if one moves towards an object, its texture expands and contracts if move away from an object. So flow of texture is invariant as it always occurs in the same way and can be considered as a direct cue of depth.

Affordances: Various cues can aid perception like

Relative brightness: Objects with brighter and clearer images are perceived as closer.

Texture gradient: Grain of textures gets smaller as the object reaches.

Relative Size: When an object moves further away from the eye, the image gets smaller. Objects with smaller images are seen as more distant.

Superimposition: If the image of one object blocks the image of another, first object is seen as closer.

Height in the visual field: Objects further away are generally higher in the visual field.

Gibson's theory seems to be based on perceiver's operating under ideal viewing conditions, where only stimulus information is sufficient. This theory could not explain that why perceptions are somewhat inaccurate, i.e. illusion. Eg. Perceptual errors like the general tendency for people to overestimate vertical extents relative to horizontal ones.

B. Richard Gregory (1970) theory of perception:

According to this theory, perception is a top-down process and can be seen as a constructive process only which involves making inferences to make a best guess about what we see.

- Perception depends on the prior knowledge and past experience only.
- Perception is a hypothesis.

Gregory explains that lot of information reaches the eyes but a proportion of them is usually lost by the time it reaches the brain. So on the basis of past experiences, brain has to guess that what a person sees. So, perception can be explained as information from sensors and past experiences and cannot be explained by a bottom-up process only. Perception changes though there is no change of sensory input, so change of appearance cannot be due to bottom-up processing.

C. Gestalt theory of Visual Perception

The Berlin School of Experimental Psychology has explained the Gestalt theory of visual perception which is based on the laws of human's skill to obtain and maintain meaningful perceptions of the world around them. The theory was proposed by the group of psychologists in 1920s in Germany to study about the perceptual organization.

The meaning of word Gestalt is organized as whole, means the individual parts have different characteristics as a whole, e.g. a tree can be explained by its parts like trunk, branches, leaves, but by looking at the tree one can aware of the overall object only, even the parts have been clearly seen they will have the secondary importance only. So basic principle of gestalt theory is that whole is different then sum of its parts.

Gestalt six principles to define perceptual organization: It means how we should combine components to perceive the whole.

- Law of Proximity: It states that things perceived as closer if they are being in sets. Principle of proximity or continuity states that things which are close together will be seen as belonging together.
- Law of Similarity: The objects that share visual characteristics like shape, size, colour, texture, orientation look alike as being grouped together.
- Law of Continuity: It predicts the preference for a continuous figure. We perceive the figure as two crossed lines instead of four lines meeting at centre.
- Law of Closure: We perceive figures with gaps in them to be complete.
- Law of Area and Symmetry: Principle of area says that in two overlapping figures, the smaller one will be perceived as a figure while the large will be observed as a ground, whereas the principle of symmetry states that the contours of a symmetrical figure isolate it from its ground.
- Law of Common Fate: If both principles of proximity and similarity are in place, then a movement takes place. They appear to change group.

4   Problem in Modelling Perception

The goal of perception is to create actions to generate new attended stimulus where any type of knowledge new or old effects processing of perception and recognition. The main aim of visual perception is to restructure the 3D world (like its geometry, topology, material, surface properties, light sources) from the observed 2D image. This visual perception is an inverse computer graphics problem and has been termed as an ill-posed problem as

- Geometry of the object is not revertible as depth range is lost.
- Reflectance, illumination, entanglement.

As human visual system is a well-developed system, it employs following different regularizing techniques to solve this highly ill-posed inverse problem efficiently and robustly

- Prior knowledge of the world
- Prior information of curve, shape, geometry
- Prior awareness of light source
- Prior understanding of materials

Our human visual system subconsciously chooses the accurate knowledge from the massive visual data in our daily life to achieve maximum efficiency and robustness.

The main problem of modelling human perception is that how to mathematically model following activities:

- How to model human perception of geometry,
- How to model human perception of depth,
- How to model human perception of light and shade.

5 Conclusion

The aim of the study is to explore the human perceptual characteristics and psychologist views for the optimization of computer vision algorithms. It can be seen that in past quantitative visual models were very famous among computer vision experts but as human visual system is a good model to imitate, we trust that perceptual models will play a substantial role in future for optimization of computer vision algorithms. Further rising volume of high-resolution video data presents new challenges and the possibility of introducing new perceptual models. It can be seen that we are still missing the feasibility to efficiently map human cognition into automated analysis process such that the manual and automated video analysis process would same results. Further the automated analysis of video system according to user's need is very difficult as to mention that which video segments are highlighting ones is a very subjective process, so obtaining the objective ground truth for the same is a very difficult task.

References

1. Desolneux A, Moisan L, Morel J-M (2004) Gestalt theory and computer vision. In: Seeing, thinking and knowing. Theory and decision library, vol 38, pp 71–101
2. Physical and social learning. http://www.visionmeetscognition.org/cogsci2015/
3. Bovik AC (2010) Perceptual video processing: seeing the future [point of view]. Proc IEEE 98(11):1799–1803
4. Majumder A (2016) Visual perception. file:///G:/Personal/Papers/Final%20Read/2-Visual%20Perception.html
5. Kato Z (2016) Computer vision. http://www.inf.u-szeged.hu/~kato/teaching/computervision/01-Introduction.pdf
6. McLeod S (2016) Visual perception theory. http://www.simplypsychology.org/perception-theories.html
7. Cognitive psychology-introduction to perception (2016). http://cognitivepsychology.wikidot.com/cognition:perception
8. Cognitive perspective of learning & information processing (2016). http://study.com/academy/lesson/cognitive-perspective-of-learning-information-processing.html

9. Computational perception and cognition. http://cvcl.mit.edu/audeoliva.html
10. Physical and social scene understanding. http://visionmeetscognition.org/cogsci2015/
11. Haque M, Murshed M (2013) Perception-inspired background subtraction. IEEE Trans Circuits Syst Video Technol 23(12):2127–2140
12. Li S, Zhang Q (2014) A computational model of visual attention based on space and object. Res J Appl Sci Eng Technol 7(1):42–48

Converting and Developing Live Web Site into a Web Content Management System

Debjani Bhowmik, Minakshi Roy, Debarshita Biswas, Suchismita Roy
and Shyera Roy

Abstract This paper analyzes the main models of collaboration in a particular existing Web site and develops a Web Content Management System (CMS). This development can help Web developers by giving overview of building and maintaining Web sites. It is faster and user friendly. It also provides easy maintenance and further updating and better facility to upload documents instantly and better appearance as well. Corporate information can be created and managed using this CMS. This management system and back-end databases allows easy updating a Web site. Content authors and marketers need the ability to change content quickly and easily. It is accessible from any computer that is connected to Internet. CMS can be used for page updating on the fly, change or upload images, and add dynamic contents to different files. It provides the developers with ready to use themes, and thus, it is a time-saving and easy operation as there is the requirement of less coding work. Here need for a CMS and system requirements to support a large content management system is described. Feedback from technical personnel also provided. It covers the complete lifecycle of the pages on site. It provides simple tools to create the content and to publish and finally to archive.

Keywords WCMS · Collaboration · Back end · Lifecycle of the page

D. Bhowmik (✉) · D. Biswas · S. Roy · S. Roy
Department of Computer Science and Engineering, The ICFAI University,
Agartala, Tripura, India
e-mail: debjani.bhowmik@iutripura.edu.in

D. Biswas
e-mail: debarshita.biswas@iutripura.edu.in

S. Roy
e-mail: suchismitaroy13atcs312@gmail.com

S. Roy
e-mail: royshreya654@gmail.com

M. Roy
Department of Computer Science and Engineering, Sikkim Manipal Institute of Technology,
Majitar, Sikkim, India
e-mail: mr9314@gmail.com

© Springer Nature Singapore Pte Ltd. 2019
H. K. D. Sarma et al. (eds.), *Advances in Communication,
Cloud, and Big Data*, Lecture Notes in Networks and Systems 31,
https://doi.org/10.1007/978-981-10-8911-4_9

79

1   Introduction

The Web site can be extremely useful for the readers as it provides quick and direct information about the Tribal Welfare Department of Tripura through the site. The main aim of the Web site is to keep the record of all the tenders, notices, and other reports related to the welfare of the scheduled tribes of Tripura and all the schemes introduced by the Government of Tripura. Web content management system along with back-end databases allows to easily updating Web site. With the help of Web content management system, authors and marketers can have the ability to change content quickly and easily. It provides responsive themes so that there is no need to write code for themes. Easy interface leads to a better understanding for the administrator, manager, and other users as well. The Web site also includes Login section for the manager of the department and Administrator and Feedback section for different user who wants to give feedback about the Web site and the department as well.

Its tools help to manage the structure of the site, the appearance of the published pages, and the navigation system becomes available for users [1]. CMS is a software program that makes building and maintaining Web sites of various types in a faster and easier way. This system helps in automatically pulling the content out and shows it on the appropriate pages based on setup rules in advance [2]. CMS allows publishing and editing, organizing, deleting contents, as well as maintenance from a central interface [3]. Such system of content management provides procedures to manage workflow in a collaborative environment. These procedures can be manual steps and automated cascade. CMSs typically aim to avoid the need for hand coding, but may help it for specific elements or entire page. All users do not have the same comfort level with technology, but the basic CMS functions and slightly more advanced ones of adding media are usually easy for everyone to grasp. In fact, anyone who can use word-processing software can use a CMS for the basic functions, thus someone no need to spend much time on training. In content management system, it contains contents that can be text document or content which can be media file such a video or image file. So, management of these contents is an important feature of every CMS [4].

In a practical business, multiple users can supply input into the Web site. Among various users, someone may add product pages to those who produce blog posts for the content marketing efforts. CMS can provide at a glance view of the status of all contents, whether it is live or being reviewed on a draft or not. It allows us to assign desired task and to check what they have done. It is also possible to integrate planned content with our marketing plan so that everyone knows what is happening, when improves site maintenance. When admin wants to change something on site then without a CMS, it can mean having travel through hundreds of pages, making changes on each one. This architecture keeps this system up to date automatically.

If the admin want to change the site of the design, CMS can make the process easy. It is possible only because the contents and design are in separate virtual boxes. So, the admin can make design changes while keeping the site functional. Mobile interface also can be added to this site. CMS not only publishes contents but also removes it when it is out of date. With a full CMS, this is very simple to publishing content. Here, all menus and links update automatically by default. So, the customers continue to have a good experience of the site. CMS can be configured to allow customized contents such as countdown calendars and lists. Instead of being reliant on external administrators, in CMS everything is under control of host admin with the ability to assign tasks and to check progress at any time whenever he desires.

2 The CMS Requirements

Web site provides interface through which administrator and manager can easily modify, manage and maintain the web site. To users/visitors, the Web site provides simple, attractive, understandable, and helpful interface through online. Multiple users can manage the Web site according to the privilege given to them. The administrator and the manager do their work according to the privilege they are provided. Different types of contents are managed and maintained easily, and if necessary new content can be added easily as per requirement. Important information or data can be easily access through the Web site because it provides all the information about the "Department of Welfare For Scheduled Tribes." In case of uploading any necessary documents or if in case there is any change, then the authenticate user or the manager can directly edit or delete, and it required very less coding and sometime no coding at all as drupal offers the features. The programmer should have the knowledge about drupal and the manager should have only the basic knowledge about drupal. Since there is no debugging tool, the programmer cannot find the error directly rather can search and rectify it by himself.

3 The Existing and the New Web Site

The new Web site provides easy means of documentation as it requires less coding. The new Web site requires less time as well as less human effort for maintenance, whereas existing Web site requires more time and human effort in order to do changes in coding. In order to attach any file, uploading any picture, uploading new version of documents, and performing any changes in designs, or any other alteration, the developer needs to change in coding which requires high programming skill, whereas in the new one, these all can happen very easily with only the knowledge of drupal.

As the new Web site is developed using drupal 7, it provides ability to create a professional looking Web site with lots of features without having to know how to write the code to make it all happen and make the Web site appearance more tendril and attractive. Drupal increases flexibility, security and also allows administrators to neatly update to new releases without customizing through overwriting. As the new Web site is developed using drupal 7 and it is a content management [5] framework, it enabled the administrator to extend the functions and utility of the core Web site through using certain modules, whereas existing one does not provide any such way of modulation.

4 Content Management System

Content management system has two main components: such as the content management application (CMA) and content delivery application (CDA). The CDA elements use and compile information of Webmaster to bring up to date the Web site. The CMA element allows the content manager or author, to manage in creation and modification of content from a Web site where expertise of a Webmaster is not essentially required. Several options are there for the application of content management system to develop a project. Depending on how advanced CMS is required and who is going to be using it, it is a troublesome to find the "perfect" CMS for a project.

4.1 Word Press

One very good example of an open resource content management system is Word Press which has almost 17% of overall sell share in Web development. Without downloading a single file, the latest version auto-update the core and plug-ins within the back end is possible.

4.2 Joomla

Joomla is unlike Drupal in which CMS is complete and it can be used for a simple portfolio site. It comes with a striking administration interface and completes with different attractive and intuitive drop-down menus. The CMS also supports access control protocols like LDAP, Open ID, and Gmail.com.

4.3 Expression Engine

Expression Engine (EE) saves very less time to understand the layout of the back end and to start creating content or modify the appearance as a whole. Expression

Engine is packed with some helpful features like the ability to have multiple sites. For designers, EE has a great engine with custom global variables, custom SQL queries and a built-in update versioning system. Template caching, query caching, and tag caching keep the site running faster.

4.4 Drupal

Drupal [6] with various elective modules can add lots of exciting features like forums, user blogs, Open ID, profiles and more. It is very easy to create a site with attractive features with a simple installation of Drupal. Actually, in Drupal with a few tertiary modules, one can create some interesting site clones with minimal effort. This application supports to build dynamic features in websites which are rich, maintainable and less time consuming.

Drupal components [6] are the fundamental system used to construct the Web site is called the Drupal core. User can add further features by adding more modules. Select themes to give a different look to the user Web site.

5 Database Connectivity

PostgreSQL is used as the database connectivity used to store the relevant data of the Web site. PostgreSQL is one of the stronger object-relational database management with an importance on extensibility and on standards compliance. It can handle workloads starting from small individual machine applications to large Internet-facing applications with many concurrent users. It handles complex SQL queries using many indexing method that are not available in other database. It has updateable views and materialized view triggers, and foreign key supports functions are stored procedures and other.

6 Analysis

Complete and accurate information requirement is essential for building a Web application. But determining the information requirement of the user is not very easy. In this project, first we have studied the entire project that we are going to complete and then we have collected some information related to Department of Welfare for Scheduled Tribes, Govt. of Tripura. After that we have designed the home page, basic pages and content types as per the requirement in developing the Web application [7]. After designing the home page and all the basic pages, we have started building content types which are required for creating content pages, to view the content pages we have used the view module.

7　Methods

The iterative waterfall model (SDLC) can be applicable to this practical Web-based development project. In this model, the various methods allow for correction of the errors committed during a phase, also when these are detected in a later phase.

7.1　System Requirements

Front End: Drupal 7.34.
Back End: Postgre SQL 9.2.
Server: Bitnami wappstack 5.4.22 apache2.
System: Compaq Presario CQ43.
Operating System: Windows 7 Ultimate.
RAM: 4 GB and higher.
Hard Disk: 500 GB and higher.
Processor: Intel Pentium(R), 2 GHz, Intel Core i3 and higher.

8　Design

The goal of design phase is to transform the requirements specified into a structure that is suitable for implementation in some programming language (Figs. 1, 2 and 3).

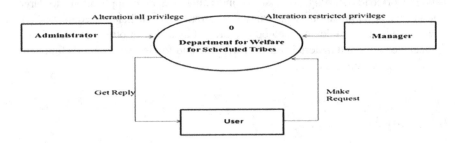

Fig. 1 Level 0

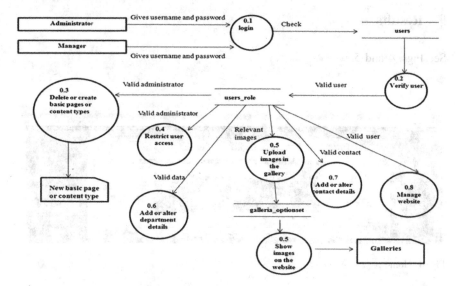

Fig. 2 Level 1

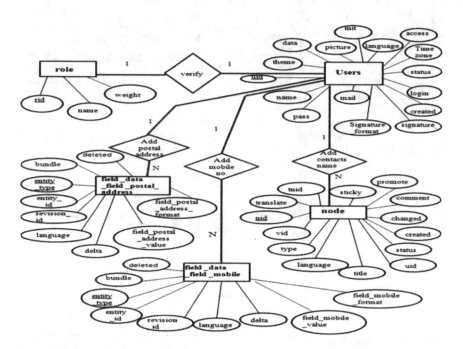

Fig. 3 E R diagram

9 Results

See Figs. 4 and 5.

Fig. 4 Home page

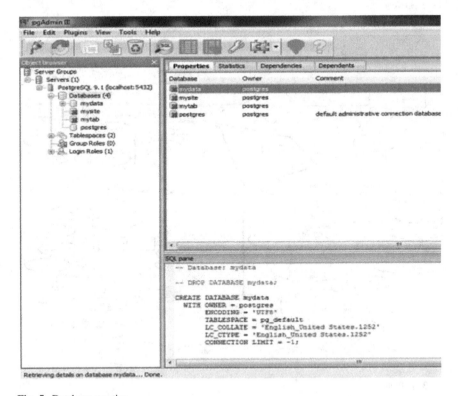

Fig. 5 Database creation

10 Conclusion

The Web site describes the information about the tribal society of Tripura. It not only describes different schemes of Tripura, but also gives information about the contact details of different welfare department officers. It gives the permission to the administrator or the manager to alter the information also provide better interface for all. The newly designed Web site provides uploading documents through content types easily rather than coding, as well as provides updating records if necessary by the assigned individual directly. It gives a proper looks of the gallery where different images welfare department are shown. It also gives the permission to the administrator or the manager to alter the information through online. The application is very much user friendly and can be handled by anyone with basic computer knowledge and English language understanding.

References

1. Shaikh MS, Fegade V (2012) Modeling essentials of content management system (CMS) for web-based MIS application. Int J Sci Technol (IJST) 1(3):379
2. Wakode BV, Chaudhari DN (2013) Study of content management systems Joomla and Drupal. IJRET: Int J Res Eng Technol. eISSN: 2319-1163, pISSN: 2321-7308
3. Sowmiya PK, Shyamala G (2014) Survey of web content management system, a collaborative environment for online. Int J Adv Res Comput Commun Eng 3(11):8504
4. Farooq A, Javed F, Hussain M, Abbas T, Hussain A (2012) Open source content management systems: a canvass. Int J Multi Sci Eng 3(10):38–43
5. Baldaniya RH, Baldaniya HJ (2014) Web development using content management system. Int J Emer Res Manag Technol 3(4). ISSN: 2278-9359
6. Nurminen JK, Wikman J, Kokkinen H, Muilu P, Gronholm M (2008, January). Drupal content management system on mobile phone. In: 2008 5th IEEE consumer communications and networking conference, IEEE, pp 1228–1229
7. Ghorechal V, Bhatt C (2013) A guide for selecting content management system for web application development. Int J Adv Res Comput Sci Manag Stud 1(3) August 2013. ISSN: 2321-7782

A Review On Existing QoS Routing Protocols in Vanet Based on Link Efficiency and Link Stability

Damodar S. Hotkar and S. R. Biradar

Abstract Vehicular ad hoc network (VANET) is a category of mobile ad hoc network (MANET). VANET provides wireless communication among vehicle-to-vehicle (V2V) and vehicle to a computing device located on roadside infrastructure called Road Side Unit (RSU). The communication between vehicles can be used for safety provisioning and emergency awareness by exchanging information among traffic participants to realize a cooperative and more efficient transportation system. An important issue associated with V2V communications used for traffic safety and emergency awareness applications (TSEAAs) is better routing with the Quality of Service (QoS) support. This paper reviews the existing QoS routing protocols based on the two important parameters: link efficiency and link stability.

Keywords MANETS · VANET · QoS · Link efficiency · Road Side Unit
Link stability · Simulation

1 Introduction

The vehicular ad hoc network (VANET) is a special class of mobile ad hoc network (MANET), where every node is a vehicle moving on a road [1]. The VANET has the potential to improve reliable communication in two ways of vehicular communication such as vehicle-to-vehicle (V-V) and vehicle-to-infrastructure (V-I) or Road Side Units (RSU). Growing use of vehicles in the city has increased the magnitude of the accidents and reduced the level of safety to the passengers [2–4]. With the proposed use of wireless technologies, development of new techniques has been encouraging to guide the driver in hazardous situations to minimize road

D. S. Hotkar (✉)
Computer Science and Engineering, VDRIT, Haliyal, India
e-mail: damodarsh418@gmail.com

S. R. Biradar
Information Science and Engineering, SDMCET, Dharwad, India

© Springer Nature Singapore Pte Ltd. 2019
H. K. D. Sarma et al. (eds.), *Advances in Communication,
Cloud, and Big Data*, Lecture Notes in Networks and Systems 31,
https://doi.org/10.1007/978-981-10-8911-4_10

dangers. However, reliable and fast delivery of safety packets to all around vehicles or RSU in a VANET at present is a significant challenge since vehicles move at different speeds. Hence, deploying reliable multi-hop routing communication either between two vehicles or vehicle–RSU by understanding the dynamics of vehicle movement in a road topology is necessary to improve the real-time safety.

Even though real-world deployment is essential to understand the effectiveness and performance of routing protocol, deploying real-world test beds involves high cost. Instead, the simulation is an effective tool to evaluate the protocols and architecture with low cost. Most of the VANET research on the design of routing protocol has relied on simulations. However, they have failed to attain the equal performance when it is to be implemented in a real-world environment. Thus, it is crucial to consider the dynamic behavior of vehicles on the road in the design of routing protocol that attempts to ensure the reliability and delay-constrained Quality of Service (QoS) requirements of real-time applications.

The VANET has the potential to improve road safety by exchanging some local information about traffic jams and accidents as instantaneously as possible. The performance of VANET routing protocols depends drastically on the link stability and quality between the communicating vehicles in terms of delay and reliability. The link stability measurement in routing plays a vital role in reliable and fast packet delivery. The mobility is affected by the drivers' plans and behaviors, and it is tough to predict accurately in link stability measurement. Recently, the mobility models include a number of other characteristics that affect the wireless communication in real life, e.g., crossroads, traffic lights, node density (traffic), and speed variations [5]. With the help of these mobility models, it is possible to represent the actual scenario of a particular area and to predict the link stability accurately. Despite better link stability, the routing protocol does not perform well in VANETs. The reason behind this is due to the less focus on other QoS attributes such as link reliability, capacity, and the geographical information between vehicles which invokes the selected paths to long delivery delay. Therefore, it is crucial to consider both the link stability and quality in a routing decision that directs the routing paths toward reliable and fast communication.

Diverse approaches have been presented in the design of QoS routing protocols in recent years. Among the existing approaches, some works mainly take advantage of QoS parameters to select the highly efficient path. These schemes consider the QoS routing as a continuous process where the routing algorithm selects the best route that satisfies the QoS constraints, and they find certain countermeasures to respond as soon as possible when the selected best route turns out to be a failure. A few works mainly focus on the stable route selection considering the mobility parameters. In order to ensure that, these works perceive vehicles' kinematic information, the mathematical distribution of their movements and velocities, and the current vehicular network conditions. However, their focus on providing multi-constrained QoS routing and trade-off between them for diverse QoS requirements under various scenarios is limited.

1.1 Significance of the QoS Routing

In the vehicular communication systems, the vehicles exchange information such as safety warnings and traffic information with one another to avoid the accidents and traffic congestion effectively. However, due to the increase in the use of vehicles day by day, the possibility of accidents is also high in recent years. A report published by the World Health Organization (WHO) shows that more than 1 million people are killed and 50 million are injured due to collisions on the roads each year. Therefore, timely and reliable communication of safety-critical data is necessary for VANET, but it is a tough task due to the lack of knowledge of the exact road map topology and dynamic driver decisions in routing. Although much mobility models take the real-time map representations, drivers' decisions, and road conditions into account, reliable delivery of safety packets to all surrounding or neighboring vehicles in a VANET is a significant challenge due to the dynamic mobility characteristics. Thus, it is important to consider the realistic mobility characteristics of each vehicle, when providing reliable service for delivering safety packets fast in highly dynamic VANETs.

1.2 Applications of QoS Routing with Realistic Mobility Model

The VANET QoS routing aims to satisfy the requirements of safety and commercial applications with the support of realistic mobility model and real road map topology. The VANETs provide a promising platform for large-scale potential applications such as control of traffic flows, blind crossing, prevention of collisions, nearby information services, and for providing Internet connectivity to the vehicular nodes while moving on the road [6].

1.2.1 Safety Applications

The safety applications include the monitoring of road, the approaching vehicles movement, the surface of the road, road curves, etc.

Traffic Control Mechanism: The VANET can provide the information of the most common traffic control mechanisms at intersections such as stop signs and traffic lights to the drivers using realistic mobility model along with the road map topology.

Interdependent Vehicular Motion: The details of the movement pattern of vehicles in realistic mobility model assist the drivers to maintain a minimum distance from the one in front of it, to increase or decrease its speed, and in the case to change over to another lane, and thus, it avoids accidents in VANET.

Post-Crash Notification: A vehicle involved in an accident broadcasts a warning message about its position to trailing vehicles so that they can change its lane and save the traveling time.

Cooperative Collision Warning: The vehicle communication alerts the two drivers potentially under crash route to mend their routes.

1.2.2 Commercial Applications

The commercial application aims at providing the traveling information and helping the drivers in controlling the vehicle.

Parking Availability: The VANET offers notification messages on the availability of parking in the urban areas, and the real road map topology assists the vehicles in locating the vacant slots in parking lots in a particular geographical area.

Internet Access: Vehicles can access the Internet through RSU when the RSU is working as a router.

Digital Map Downloading: The drivers can download a map of the regions for travel guidance before traveling to a new area. Content Map Database Download is used to get this valuable information from mobile hotspots.

Several works have been carried on the field of QoS routing in VANETs. This paper deals with the survey of such very interesting and thought-provoking researches which has made to put forward the idea of real-time vehicular communications for efficient and reliable QoS. This paper lists brief description of the works that are conducted on the above-mentioned aspects of vehicular communications.

2   Related Works

The VANETs support a variety of applications for safety and driving comfort. Therefore, guaranteeing a stable and efficient QoS routing over VANETs is important and essential in real-time vehicular communications. Efficient and reliable QoS-based routing protocols [7, 8] are crucial to forward data packets within the required QoS constraints in the targeted regions and also to be adaptable well with various applications. Most of the researchers have designed several QoS routing protocols for VANETs. The existing QoS routing protocols are classified into two types such as the link efficiency and link stability.

2.1   Link Quality-Based Routing

A traffic-aware intersection-based geographical routing protocol, TIGeR in [9] exploits only the nodes at intersections to make a routing decision based on vehicular traffic information on diverse roads and the road's angle on the destination. The work in [10] presents an improved Greedy Traffic-Aware Routing Protocol (GyTAR) that exploits the intersection geographical information to determine robust and optimal routes within the urban environments. An AntRS is a multi-objective Heuristic Algorithm Based on Ant Colony Optimization for Multi-objective Routing [11]. The study of AntRS states that in a real VANET network, it finds alternative paths of communication when the obstacles fade the signal transmitted and hinder the communication. An Intersection-based Delay Sensitive Routing using Ant colony optimization (IDRA) for VANETs in urban environments has been proposed in [12]. The IDRA computes the optimal route between two terminal intersections and the route closest to the source and the destination vehicles. A novel adaptive multi-criteria VANET routing protocol, named as vehicular routing protocol based on Ant Colony Optimization (VACO), aims to determine the best routes from a source to the destination with quality metrics estimated regarding latency, bandwidth, and delivery ratio [13]. The works [12, 13] only focus on path quality and they lack in considering the path stability QoS metrics.

A Situation-Aware Multi-constrained QoS (SAMQ) routing algorithm has been developed in [14] which computes feasible routes for vehicular communication subject to multiple QoS constraints. The SAMQ algorithm considers the Situation-Aware (SA) and exploits Ant Colony System (ACS) with the aim of assuring a reliable data transmission in VANETs. An Adaptive Routing Protocol based on QoS and vehicular Density (ARP-QD) [15] is an intersection-based routing protocol that finds the best path for end-to-end data delivery. The ARP-QD considers diverse QoS requirements, the link duration, and connectivity in routing path selection. It maintains the trade-off between stability and efficiency. The ARP-QD obtains neighbors' information based on local vehicular density to reduce the network overhead. Even though, connectivity estimation using beacon packets dissemination causes channel congestion in the high traffic environment. The instantaneous value of the neighbor discovery algorithm is not accurate in dynamic scenarios. Also, using global distance alone is not adequate to reflect the complete QoS of a routing path as the data packets are likely to suffer from network congestion in the upcoming road segments.

2.2 Link Stability-Based Routing

One of the simple and efficient methods to escalate the link stability is to determine the longest link duration. The work in [16] designs a Receive On Most Stable Group-Path (ROMSGP) algorithm that selects the most stable path based on the link expiration time. The ROMSGP does not consider any QoS parameters in path selection. As a result, it may not be suitable for emergency scenarios. The main objective of [17, 18] is to identify the high stable routing path and avoid carry-and-forward delay. The links with good connectivity have the least distance that invokes the selected stable paths to include more hops and results in longer data delivery delay. A stable VANET routing protocol provides fast and reliable data delivery based on the real-time road vehicular density [19]. However, the real-time update of density vehicle information sustains a large amount of communication overhead, which results in its performance deterioration in large-scale VANETs.

The work in [20] utilizes the vehicle's speed and position to identify relatively stable links. There is need to trade adaptively off the path efficiency and path stability for diverse QoS requirements in diverse applications. An intersection-based geographical routing protocol in [21] aims to discover the routing path with high connectivity probability and other QoS constraints. However, the previous research mainly focuses on the link connectivity and sparingly uses the geographical distance information among vehicles. As a result, the selected paths may have additional loops that cause longer data delivery delay. Thus, it is not compatible with the applications which require a shorter delivery delay.

2.3 Mobility Modeling and Prediction Techniques for Realistic Routing

The fuzzy-assisted social-based routing (FAST) protocol in [22] considers the advantage of the social behavior of drivers on the road in making the routing decision. It leverages the friendship mechanism according to the global knowledge of vehicular traffic information to make critical decisions at intersections. The work in [23] proposes two urban traffic amount prediction models based on the propagation of traffic flow and the spare road capacity, respectively. It applies the traffic amount prediction models to a route guidance system (RGS) and reduces the average travel time. A Connectivity-aware Routing (CAR) in [24] utilizes realistic mobility traces and road maps in route estimation, and it forwards the data packets along the founded path. Then vehicle-assisted data delivery (VADD) in [25] uses a snapshot of a real street map, and it derives the street layout. The VADD follows the predictable vehicle mobility in route selection to forward the packets according to existing traffic pattern. In [26], an EASE aims to address the problems of routing in large-scale mobile ad hoc networks. The EASE exploits the movement of vehicles, and it encounters the mobility history with the other routes used. A Prediction-based Soft Routing Protocol (PSR) in [27] utilizes the real traces that are collected from more than 4,000 taxies at Shanghai over six months to identify the Vehicular Mobility Pattern (VMP) through the Variable-order Markov (VOM) model scheme. The picking up of the mobility model parameters alone in route computation cannot guarantee a reliable data transmission, as it does not satisfy certain QoS requirements such as drivers' dynamic decisions according to traveling environment.

3 Conclusion

In this paper, we have reviewed the related works on link efficiency and link stability and concluded that most of the conventional QoS routing approaches in VANET focus on either link efficiency or stability. It is essential to focus on providing multi-constrained QoS routing and trade-off between them for diverse QoS requirements in different scenarios.

References

1. Zeadally S, Hunt R, Chen YS, Irwin A, Hassan A (2012) Vehicular ad hoc networks (VANETS): status, results, and challenges. Telecommun Syst 50(4):217–241
2. Chen R, Jin W-L, Regan A (2010) Broadcasting safety information in vehicular networks: issues and approaches. IEEE Netw 24(1):20–25
3. Chen W, Cai S (2005) Ad hoc peer-to-peer network architecture for vehicle safety communications. IEEE Commun 43(4):100–107
4. Torrent-Moreno M, Mittag J, Santi P, Hartenstein H (2009) Vehicle to-vehicle communication: fair transmit power control for safety-critical information. IEEE Trans Vehic Technol 58(7):3684–3703
5. Chan WF, Sim ML, Lee SW (2007) Performance analysis of vehicular ad hoc networks with realistic mobility pattern. In: Proceeding IEEE international conference telecommunications, pp 318–323
6. Tomer P, Chandra M (2010) An application of routing protocols for vehicular ad-hoc networks. In: International conference on networking and information technology (ICNIT), pp. 157–160
7. Yan G, Rawat D, Bista B (2012) Provisioning vehicular ad hoc networks with quality of service. Int J Space Based Situated Comput 2(2):104–111
8. Bitam S, Mellouk A (2011) QoS swarm bee routing protocol for vehicular ad hoc networks. In: Process IEEE international conference on communications, pp 1–5
9. Tavakoli R, Nabi M (2013) TIGeR: a traffic-aware intersection-based geographical routing protocol for urban VANETs. In: Vehicular technology conference (VTC Spring) IEEE, pp 1–5
10. Jerbi M, Senouci S-M, Rasheed T, Ghamri-Doudane Y (2009) Towards efficient geographic routing in urban vehicular networks. IEEE Trans Veh Technol 58(9):5048–5059
11. Silva R, Lopes HS, Godoy W (2013) A heuristic algorithm based on ant colony optimization for multi-objective routing in vehicle ad hoc networks. In: Computational intelligence and 11th Brazilian congress on computational intelligence, pp 435–440
12. Li G, Boukhatem L (2014) An intersection-based delay sensitive routing for VANETs using ACO algorithm. In: Process IEEE international conference on computer communications and networks, pp 750–757
13. Li G, Boukhatem L (2013) Adaptive vehicular routing protocol based on ant colony optimization. In: Process IEEE international workshop on vehicular ad hoc networks, pp 95–98
14. Eiza MH, Owens T, Ni Q, Shi Q (2015) Situation-aware QoS routing algorithm for vehicular ad hoc networks. Veh Technol IEEE Trans 64(12):5520–5535
15. Sun Y, Luo S, Dai Q, Ji Y (2015) An adaptive routing protocol based on QoS and vehicular density in urban VANETs. Int J Distrib Sens Netw 11:5
16. Taleb T, Sakhaee E, Jamalipour A, Hashimoto K, Kato N, Nemoto Y (2007) A stable routing protocol to support its services in VANET networks. IEEE Trans Veh Technol 56(6):3337–3347

17. Rezende C, Pazzi RW, Boukerche A (2010) Enhancing path stability towards the provision of multimedia support in vehicular ad hoc networks. In: Proceedings of the IEEE international conference on communications (ICC'10), pp 1–5
18. Yang Q, Lim A, Agrawal P (2008) Connectivity aware routing in vehicular networks. In: Proceedings of the IEEE wireless communications and networking conference (WCNC'08), pp 2218–2223
19. Yu H, Ahn S, Yoo J (2013) A stable routing protocol for vehicles in urban environments. Int J Distrib Sens Netw 9:759361
20. Huang H, Zhang S (2013) A routing algorithm based on dynamic forecast of vehicle speed and position in vanet. Int J Distrib Sens Netw 9:390795
21. Saleet H, Langar R, Naik K, Boutaba R, Nayak A, Goel N (2011) Intersection-based geographical routing protocol for VANETs: a proposal and analysis. IEEE Trans Veh Technol 60(9):4560–4574
22. Khokhar RH, Noor RM, Ghafoor KZ, Ke CH, Ngadi MA (2011) Fuzzy-assisted social-based routing for urban vehicular environments. EURASIP J Wirel Commun Netw 1:1–15
23. Liang Zilu, Wakahara Yasushi (2014) Real-time urban traffic amount prediction models for dynamic route guidance systems. EURASIP J Wirel Commun Netw 1:1–13
24. Naumov V, Gross TR (2007) Connectivity-aware routing (CAR) in vehicular ad hoc networks. In: Proceedings of the 26th IEEE international conference on computer communications (INFOCOM), pp 1919–1927
25. Zhao J, CaoG (2006) VADD: vehicle-assisted data delivery in vehicular ad hoc networks. In: Proceedings of the 25th IEEE international conference on computer communications, pp 1–12
26. Grossglauser M, Vetterli M (2006) Locating mobile nodes with ease: learning efficient routes from encounter histories alone. IEEE/ACM Trans Netw 14(3):457–469
27. Xue Guangtao, Luo Yuan, Jiadi Yu, Li Minglu (2012) A novel vehicular location prediction based on mobility patterns for routing in urban VANET. EURASIP J Wirel Commun Netw 1:1–14

A Diamond-Shaped Fractal Bow-Tie Antenna for THz Applications

Malay Ranjan Tripathy, Vipin Choudhary, Aastha Gupta, Priya Ranjan and Daniel Ronnow

Abstract A compact multiband diamond-shaped fractal bow-tie terahertz antenna is designed on FR4_epoxy substrate with permittivity 4.4. The dimension of substrate is $4 \times 6 \times 1.5$ mm^3. Multi-bands are obtained in S_{11} versus frequency plots. The effects of creating defects in the ground plane of the antenna were studied and analyzed to get the optimized performance of antenna in the frequency bands at 0.500, 0.560, and 0.594 THz, respectively.

Keywords THz antenna · Bow-tie · Multiband · Ground defect surface

1 Introduction

THz antenna research is becoming more interesting because of the unique features of THz waves [1–3]. THz technology can be used in various applications such as ultrahigh speed wireless communication, high-resolution imaging, high-speed spectroscopy, security control, and space communication [4–8]. THz radiation is non-ionizing, non-destructive, and hence relatively safe. It can produce results better than X-rays [9]. Various efforts are being made to investigate different antenna geometries for better bandwidth and gain. Some of them are THz graphene antenna, on-chip antennas, leaky wave antennas, spiral antennas, and butterfly antennas [10–15]. Furthermore, by incorporating special designs of these antennas

M. R. Tripathy (✉) · V. Choudhary · A. Gupta
Department of Electronics and Communication Engineering,
ASET, Amity University Uttar Pradesh, Noida, Uttar Pradesh, India
e-mail: mrtripathy@amity.edu

M. R. Tripathy · D. Ronnow
Department of Electronics, Mathematics and Natural Sciences,
University of Gavle, Gavle, Sweden

P. Ranjan
Department of Electrical and Electronics Engineering,
ASET, Amity University Uttar Pradesh, Noida, Uttar Pradesh, India

© Springer Nature Singapore Pte Ltd. 2019
H. K. D. Sarma et al. (eds.), *Advances in Communication,
Cloud, and Big Data*, Lecture Notes in Networks and Systems 31,
https://doi.org/10.1007/978-981-10-8911-4_11

or bow-tie plasmonic antennas and metamaterial antennas, the radiation characteristic of THz antennas has increased tremendously. Recently, it has been realized
that the radiation features of a bow-tie THz antenna can be improved by incorporating a fractal structure of spatial distribution of surface currents [16].

Fractal antennas have become attractive because of their small size compared to
the wavelength at the operational frequency which is due to the self-similar and
space-filling geometrical structures. An object composed of smaller copies of it is
called a self-similar or fractal structure. This was introduced for the first time by
Mandelbrot [17] as "a shape made of parts similar to the whole in some way." The
definition of self-similarity in more detail is found in [18]. Different types of fractals
are reported in the literature such as Sierpinski antenna, Minkowski dipole, Hilbert
curve, Koch curves, and fractal trees. A review on this is made in the paper [19].
Fractal antennas are used in various applications [19–21]. Further, the mixture of
conventional design with fractal geometries is made more suitable for higher-end
applications because of its better directivity, gain, and multiband features [16, 21].

However, THz technology has the limitation to have signal generation in the
challenging range of 100 GHz and above. Similarly, THz antenna characteristics
such as bandwidth, impedance, polarization are needed to be better understood and
therefore continue to be a major research challenge. Realization of ultra-broadband
antenna, very large antenna array, high-speed architecture, propagation modeling,
etc., have attracted much attention in recent days [1, 3, 22]. In addition, it is
interesting for researchers to understand the way THz antennas interact with
electromagnetic waves, its fabrication, signal generation, and detection.

This paper proposes a diamond-shaped fractal bow-tie antenna for THz application. Stepped impedance feeding and defected ground surface are used to improve the
impedance matching of the design. Different iterations are made, and return loss,
bandwidth, and gain of antenna are compared, studied, and analyzed. FR4 substrate is
used for this design. The simulation is carried out by using HFSS software.

The paper is organized as follows. Section 2 describes antenna design and characteristics. Results and discussion are made in Sect. 3. Conclusion is made in Sect. 4.

2 Antenna Design

The proposed antenna has used FR4_epoxy as substrate with a dielectric constant of
4.4. It is one of the popular substrates for the antenna fabrication. The electrical
properties of this material such as dielectric permittivity and loss tangent are frequency dependent. Beziuk et al. [23] have studied these parameters and found small
variation in the THz frequency range. The dimension of the substrate is used as
$4 \times 6 \times 1.5$ mm^3. The ground plane dimension is 2×6 mm^2. In this design, two
symmetric diamond-shaped patches were created. Different iterations were made to
find the fractal shape. These fractal shapes are combined with the conventional
bow-tie structure to create a diamond-shaped fractal bow-tie antenna. Stepped
impedance microstrip line feeding is used to improve the impedance matching.

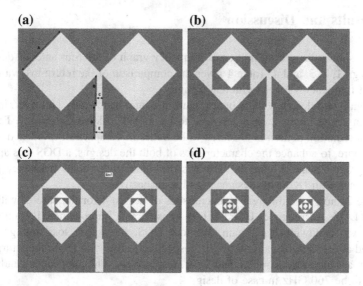

Fig. 1 **a** Design of antenna in zeroth iteration (design 1), **b** first iteration (design 2), **c** second iteration (design 3), and **d** third iteration (design 4)

The ground plane is defected and optimized for better return loss, bandwidth, and gain. Figure 1a shows the proposed antenna at the zeroth iteration. The dimensions of the antenna are given as $A = 1.97$, $B = 1.1$, $C = 0.2$, $D = 1.2$, and $E = 0.3$ mm.

In the first iteration, a rectangular slot is taken out from the diamond patch and a new diamond shape of side 0.77 mm is formed within the rectangular slot. Figure 1b shows the antenna after the first iteration. Figure 1c shows the second iteration with another rectangular slot with the self-similar structure of earlier. The third iteration is shown in Fig. 1d. The design in zeroth, first, second, and third iterations are referred to as design 1, design 2, design 3, and design 4, respectively.

In order to achieve the enhanced characteristics, defective ground structure (DGS) technique was used. The defect in the ground metal plane is created which shows the considerable effects on the characteristics of the proposed antenna. Figure 2a, b shows the ground (back) view of the antenna designs 3 and 4 without and with DGS, respectively. Introduction of DGS increased the gain characteristic considerably.

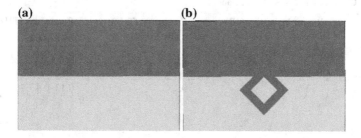

Fig. 2 Back view of designs 3 and 4 **a** without DGS and **b** with DGS

3   Results and Discussion

Figure 3 shows the return loss versus frequency graph of the four antenna designs, i.e., design 1, 2, 3, and 4. Table 1 gives the comparison of the return loss values at different frequencies for the different designs.

Design 3 and design 4 show better return loss and bandwidth. However, in case of design 4, the frequency band is seen to be shifted to lower frequency. But the gain and radiation pattern is observed to be better in comparison to other designs. Furthermore, to enhance the characteristics of both the designs, a DGS was applied on the ground plane. Figure 4a, b shows the return loss versus frequency graph without and with DGS, for designs 3 and 4, respectively.

Table 2 shows the different characteristics of design 4 for the case of with and without DGS. At frequency 0.560 THz, a considerable improvement is seen in the return loss, bandwidth, and gain. Without DGS, the return loss and gain are observed as −13.89 dB gain of 2.7 dB, respectively. But with DGS, the return loss and gain are increased to −22.1 and 9 dB, respectively. The impedance bandwidth is seen to be 760 GHz in case of design 4.

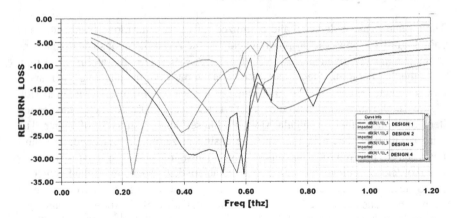

Fig. 3   Return loss versus frequency graph of designs A, B, C, and D

Table 1   Comparison design 1, 2, 3, and 4

Designs	Return loss (S_{11}) (dB)	Frequency (THz)	Bandwidth (THz)
1	−32.94	0.595	0.18–0.69
2	−24	0.39	0.23–0.6
3	−38.5	0.45	0.25–0.7
4	−33.5	0.233	0.13–0.4

(a)   (b)

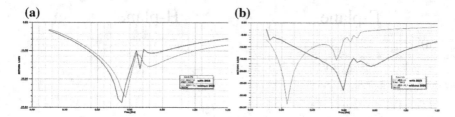

Fig. 4 Comparison of return loss without and with DGS for **a** design 3 and **b** design 4

Table 2 Resultant characteristics of design 4 without and with application of DGS

DGS	Freq (THz)	S_{11} (dB)	Gain (dB)	S_{11}(min) (dB)	F_L–F_H (THz)
Not applied	0.500	−9.27	0.5	−33.3 at	0.13–0.4
	0.560	−13.89	2.7	0.233 THz	0.5–0.57
	0.594	−7.25	1	−15.3 at	
				0.54 THz	
Applied	0.500	−17.3	13	−27.75 at	0.28–1.0
	0.560	−22.1	9	0.594 THz	
	0.594	−27.75	10.2		

Figure 5 shows the gain versus frequency graph of design 4. It is seen that the gain is better in case of the design with DGS in comparison to without DGS. The peak gain of design with DGS is 7 dB higher than the design without DGS. The gain is more than 5 dB throughout the frequency sweep in case of design with DGS.

Figure 6a–c shows E-plane and H-plane radiation patterns for design 4 at frequencies of 0.500, 0.560, and 0.594 THz, respectively. It is seen that omnidirectional radiation patterns are obtained in all E- and H-planes. It is advantageous to have omnidirectional radiation pattern in many communication applications. In case of the E-plane, the radiation patterns for designs with DGS have better gain in comparison to design without DGS. It shows the effect of DGS on the antenna

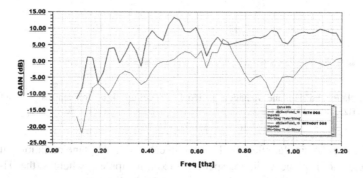

Fig. 5 Gain versus frequency graph for design D

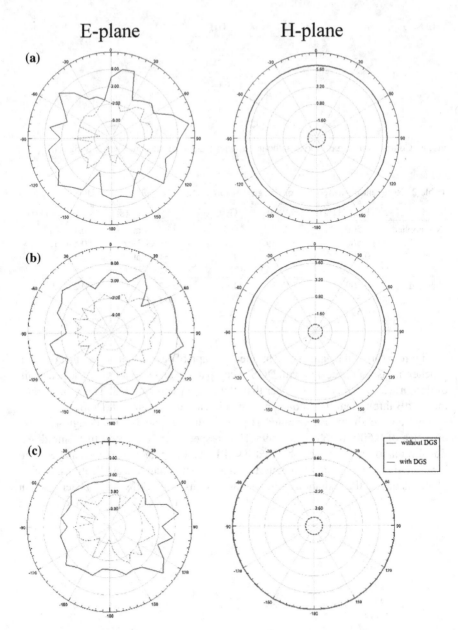

Fig. 6 E-plane and H-plane radiation patterns for design 4 at frequencies **a** 0.500 THz, **b** 0.560 THz, and **c** 0.594 THz

characteristics. A similar effect is also seen in case of the H-plane radiation patterns. However, the E-plane radiation patterns exhibit ripples, whereas the H-plane radiation patterns have less gain with more stable radiations.

4 Conclusion

In this paper, the diamond-shaped fractal bow-tie antenna is proposed for THz application. Stepped impedance microstrip line feeding and DGS techniques are used to improve the antenna parameters. The different results of antenna with and without DGS are compared and analyzed. The effects of DGS on the improvement of antenna features are observed in this investigation. The peak gain is obtained as 13 dB and the maximum bandwidth is seen to have 760 GHz. The radiation patterns are found to be omnidirectional. It is good to have omnidirectional radiation patterns for the communication. This antenna can be used for wide applications in THz range.

References

1. Kleine-Ostmann Thomas, Nagatsuma Tadao (2011) A review on terahertz communications research. J Infrared Millimeter Terahertz Waves 32(2):143–171
2. Tonouchi Masayoshi (2007) Cutting-edge terahertz technology. Nat Photonics 1(2):97–105
3. Akyildiz IF, Jornet JM, Han C (2014) Terahertz band: next frontier for wireless communications. Phy Commun 12:16–32
4. Koenig S, Lopez-Diaz D, Antes J, Boes F, Henneberger R, Leuther A, Tessmann A (2013) Wireless sub-THz communication system with high data rate. Nat Photonics 7(12):977–981
5. Trichopoulos GC, Mosbacker HL, Burdette D, Sertel K (2013) A broadband focal plane array camera for real-time THz imaging applications. IEEE Trans Antennas Propag 61(4): 1733–1740
6. Kim JY, Song HJ, Ajito K, Yaita M, Kukutsu N (2013) Continuous-wave THz homodyne spectroscopy and imaging system with electro-optical phase modulation for high dynamic range. IEEE Trans Terahertz Sci Technol 3(2):158–164
7. Corsi C, Sizov F (2014) THz and security applications: detectors, sources and associated electronics for THz applications. Springer, New York
8. Choudhury B, Sonde AR, Jha RM (2016) Terahertz antenna technology for space applications. Terahertz antenna technology for space applications. Springer, Singapore, pp 1–33
9. Sun Y, Sy MY, Wang YX, Ahuja AT, Zhang YT, Pickwell-MacPherson E (2011) A promising diagnostic method: terahertz pulsed imaging and spectroscopy. World J Radiol 3(3):55–65
10. Jornet JM, Akyildiz IF (2013) Graphene-based plasmonic nano-antenna for terahertz band communication in nanonetworks. IEEE J Sel Areas Commun 31(12):685–694
11. Correas-Serrano Diego, Gomez-Diaz Juan Sebastian, Alù Andrea, Alvarez-Melcon Alejandro (2015) Electrically and magnetically biased graphene-based cylindrical waveguides: analysis and applications as reconfigurable antennas. IEEE Trans Terahertz Sci Technol 5(6):951–960
12. Klein B, Seiler P, Plettemeier D (2015) On-chip fractal bowtie-antenna for 185 GHz to 200 GHz. In: IEEE International symposium on antennas and propagation and USNC/URSI national radio science meeting, 2015 IEEE, pp 1452–1453
13. Esquius-Morote M, Gómez-Dı JS, Perruisseau-Carrier J (2014) Sinusoidally modulated graphene leaky-wave antenna for electronic beamscanning at THz. IEEE Trans Terahertz Sci Technol 4(1):116–122

14. Berry CW, Hashemi MR, Jarrahi M (2014) Generation of high power pulsed terahertz radiation using a plasmonic photoconductive emitter array with logarithmic spiral antennas. Appl Phys Lett 104(8):081122–081124

15. Trichopoulos GC, Mosbacker HL, Burdette D, Sertel K (2013) A broadband focal plane array camera for real-time THz imaging applications. IEEE Trans Antennas Propag 61(4): 1733–1740

16. Zhong YW, Yang GM, Zhong LR (2015) Gain enhancement of bow-tie antenna using fractal wideband artificial magnetic conductor ground. Electron Lett 51(4):315–317

17. Mandelbrot BB (1983) The fractal geometry of nature. W. H. Freeman and Company, New York

18. Peitgen HJHO, Saupe D (1990) Chaos and fractals: new frontiers of science. Springer, New York

19. Cohen N (2015) Fractal antenna and fractal resonator primer. Benoit Mandelbrot: A Life in Many Dimensions. World Scientific, Singapore, pp 207–228

20. Puente JRC, Cardama A (1999) Fractal-shaped antennas. In: Werner RMDH (ed) Frontiers in electromagnetics. Wiley-IEEE Press, Hoboken, NJ, pp 48–93

21. Werner DH, Ganguly S (2003) An overview of fractal antenna engineering research. IEEE Antennas Propag Mag 45(1):38–57

22. Song HJ, Nagatsuma T (2011) Present and future of Terahertz communications. IEEE Trans Terahertz Sci Technol 1(1):256–263

23. Beziuk G, Jarzab PP, Nowak K, Plinski EF, Walczakowski MJ, Witkowski JS (2012) Dielectric properties of the FR-4 substrates in the THz frequency range. In: 37th international conference on infrared, millimeter, and terahertz waves (IRMMW-THz), Wollongong, NSW, (2012), pp 1–2

A Study on Few Approaches to Counter Security Breaches in MANETs

Moirangthem Goldie Meitei and Biswaraj Sen

Abstract Mobile ad hoc networks (MANETs) represent a class of networking that is quite essential and different from the traditional systems. Though the use of MANETs is gaining popularity in academic and commercial domains, MANETs have been initially designed to be deployed in areas such as emergency search and rescue operations, military battlefields, and other hostile or challenging environments. Because of the demanding environments that they have to operate in, MANETs do not have well-defined infrastructure unlike wired networks. All the participating nodes in a MANET work via cooperation and hence central coordination is absent. This places an inherent trust among the nodes forming the network in a MANET. Another major consideration regarding MANETs is that they have to often deal with limited resources such as power and bandwidth. These characteristic properties of MANETs make them susceptible to different kinds of attacks which aim to find vulnerabilities in the MANET protocols or target the limited resources. Hence, it becomes essential to recognize these threats and find ways to mitigate and tackle them. This paper will emphasize in understanding threats in ad hoc networks and the approaches to deal with these threats.

Keywords MANET · Intrusion detection · Trust · Software agent

1 Introduction

Ad hoc networks, as the term ad hoc suggests, refer to wireless networks that are constructed for a particular purpose or an immediate need. Ad hoc networks differ from traditional networking systems in that they do not require a centralized

M. G. Meitei (✉) · B. Sen
Computer Science and Engineering Department, Sikkim Manipal Institute
of Technology, Majitar, Sikkim, India
e-mail: mgmeitei@gmail.com

B. Sen
e-mail: biswaraj.s@smit.smu.edu.in

© Springer Nature Singapore Pte Ltd. 2019
H. K. D. Sarma et al. (eds.), *Advances in Communication,
Cloud, and Big Data*, Lecture Notes in Networks and Systems 31,
https://doi.org/10.1007/978-981-10-8911-4_12

coordinator or prior infrastructure to be in place. Thus, ad hoc networks are also called infrastructure less networks [1]. Such networks use a wireless medium for communication. A mobile ad hoc network (MANET) refers to a network in which the nodes forming the ad hoc network are mobile [2].

In a MANET, the nodes cooperate with each other to share information. If a destination node falls beyond the transmission range of a node that wants to transmit information, the sender node transmits the information to its neighbor which in turn propagates it to its neighbors until it reaches the required destination. It can be seen that this infrastructure places an inherent trust in all other nodes in the network for information propagation. A malicious attacker can take advantage of this trust relationship among the nodes, thereby compromising the network. Also due to the mobility of the nodes and the dynamically changing network topology, it is hard to determine if a packet is getting dropped because of the intrinsic network characteristics or because of the presence of a malicious attacker in the network. Hence, care must be taken when detecting threats that the generation of false alarms should be minimal.

This paper briefly discusses the different types of attacks that can take place in a MANET and the different strategies that can be used to tackle the security threats. The rest of the paper is organized as follows: Sect. 2 discusses the various security threats in MANETs. Section 3 explores some of the different mechanisms that have been proposed to tackle some of the security threats. Section 4 gives a brief summary of the attacks and security mechanisms, and Sect. 5 provides the conclusion.

2   Security Threats in MANETs

Security is a very important aspect in MANETs, especially since ad hoc networks are deployed in hostile environments such as military battlefields. The task of routing in MANETs faces several challenges because of its innate network characteristics and the areas of deployment. Some of these challenges are as follows [3]:

(a) Mobility
(b) Bandwidth constraint
(c) Error-prone and shared channel
(d) Hidden and exposed terminal problems
(e) Location-dependent contention
(f) Other resource constraints such as computing power, battery power, buffer storage, etc.

MANETs face vulnerabilities because of shared wireless medium, lack of physical protection for the mobile nodes, and complete trust among nodes because of lack of centralized decision-making entity [4]. MANETs operate by establishing an inherent trust relationship among its participating nodes. Hence, each node in a MANET is able to function as a router. But since the wireless medium is shared and there is a lack of central coordination, MANETs are vulnerable to attacks from other devices within the transmission range. Thus, managing trust also becomes an important issue [5].

Also MANETs lack a clear line of defense since there is no well-defined place where traffic monitoring or access control mechanisms can be deployed [6]. Although cryptography can be used to provide security services such as confidentiality, authentication, integrity, and non-repudiation, it is not sufficient to deal with attacks that compromise on availability, such as DoS attacks [7]. MANETs face different kinds of security threats as follows:

(a) Denial of service
(b) Resource consumption in the form of energy depletion and buffer overflow
(c) Host impersonation
(d) Information disclosure
(e) Interference

The attacks against mobile ad hoc wireless networks can be generally classified into two types [1]:

(a) Passive attacks

Passive attacks are those attacks in which the malicious nodes do not disrupt the working of the network, but they listen to the data being transferred without altering it. These kind of attacks can violate the confidentiality of the data being sent in the network.

Some examples of passive attacks are eavesdropping, traffic analysis, and monitoring. These attacks are associated with the Physical layer and Link layer [7].

(b) Active attacks.

Active attacks, on the other hand, are those attacks that disrupt the working of the network by either altering or destroying the data. These attacks can be divided into two types as follows:

- External attacks: These are the attacks that are performed by nodes that do not belong to the network.
- Internal attacks: These are the attacks that are performed by nodes from within the network.

Examples of active attacks include jamming, spoofing, modification, replaying, DoS. These attacks are associated with Physical layer, Network layer or across multi-layers [7].

Some of these attacks in MANETs are discussed as follows:

Eavesdropping:

The act of intercepting messages and reading them by unauthorized attackers without actually modifying the messages is known as eavesdropping. In MANETs, the mobile nodes share a wireless medium in which messages are usually broadcast over the network. These broadcast messages over the wireless medium can be easily intercepted by tuning to the particular frequency of the message.

Black hole attack:

A black hole attack is a DoS attack in which a malicious node falsely claims that it has the shortest path to the destination node. It is an active attack type which targets vulnerabilities in on-demand routing protocols such as DSR and AODV (Fig. 1).

Black hole attack is carried out by an attacker by sending fake routing information [8]. In this attack, an attacker node first claims that it has the shortest route to a given destination when it receives Route Request message from a sender node. For this, the attacker replies to the Route Request message with a Route Reply having a very high destination sequence number, hence ensuring that the attacker gets included in the route from the sender to the destination. On receiving the subsequent data packet from the sender, the attacker will not forward the data packets but instead drop them, thus preventing them from reaching the intended destination. A subtler version of Black hole attack can selectively forward data packets which makes it even harder to detect the attacker.

Gray hole attack:

A gray hole attack is an active attack type which causes dropping of messages. It can also be considered as a variant of Black hole attack. In this attack, the attacking node first honestly replies to Route Request message with correct Route Reply message. Then when the attacker receives data packets to be sent to the destination, it drops either some or all of the data packets intended for the destination node. Gray hole attacks are harder to detect than black hole attacks because it is difficult to conclude whether the packets are being dropped intentionally or because of a genuine network congestion.

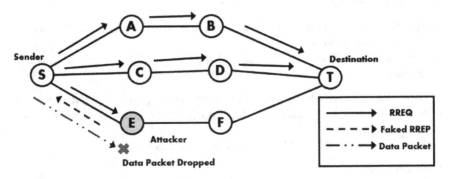

Fig. 1 Black hole attack

Rushing attack:

In rushing attack, a malicious node which receives a Route Request packet from the source node floods the packet quickly throughout the network before other nodes which also receive the same Route Request packet can react [3]. This is possible because the malicious attacker ignores the delays imposed by the network protocol. Nodes that receive the legitimate Route Request packets assume those packets to be duplicates of the packet already received through the adversary node and hence discard those packets. Any route discovered by the source node would contain the malicious node as one of the intermediate nodes. Hence, the source node would not be able to find secure routes, that is, routes that do not include the malicious node (Fig. 2).

Sleep deprivation

Sleep deprivation is a resource consumption attack which attacks the limited battery life of a MANET node. In this attack, an attacker or a compromised node can attempt to consume battery life by requesting excessive route discovery, or by forwarding unnecessary packets to the victim node [7]. This leads to exhaustion of battery life of the node, thus compromising the performance of the network.

Wormhole

A wormhole attack is an attack carried out by two colluding attackers in the network. In this attack, an attacker receives packets at one point in the network, "tunnels" them to another point in the network, and then replays them into the network from that point [9]. Often, the colluding attackers are connected by a private high-speed network which provides faster transmission than the wireless medium of the network (Fig. 3).

In ad hoc routing protocols such as AODV and DSR, a wormhole attack may be launched in such a way that an attacker receiving a Route Request message will forward it to its colluding attacker who in turn rebroadcasts the request to its neighbors. The neighbors will discard all subsequent Route Requests thinking them

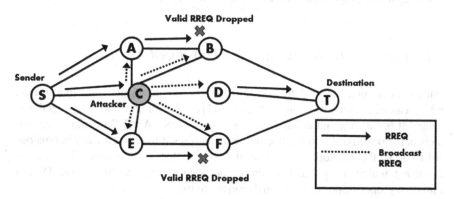

Fig. 2 Rushing attack

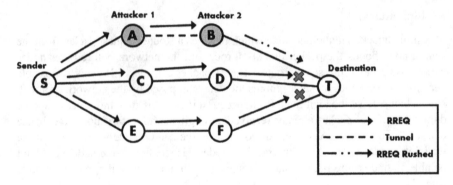

Fig. 3 Wormhole attack

to be duplicates. Thus, this prevents all other routes to the destination except the one containing the colluding attackers.

It can be seen that because of the vulnerabilities of MANETs, attacks can take place in several layers. However, this paper will look at security measures for attacks in the Network layer only.

3 Mechanisms for Dealing with Security Threats

Many scholars have proposed several types of defense mechanisms for dealing with the security threats mentioned in the previous section. Some of these defense mechanisms that are being addressed by this paper are:

(a) Intrusion Detection System (IDS)-based approach
(b) Authentication-based approach
(c) Software agent-based approach

These approaches are further discussed as follows.

3.1 Intrusion Detection System (IDS)-Based Approach

Intrusion detection can be defined as a process of monitoring activities in a system, which can be a computer or network system [10]. The mechanism by which this is achieved is called an intrusion detection system (IDS). An IDS monitors and collects network activity information and then analyzes it to check for any anomalous behavior in the network. If an IDS determines that an anomalous behavior is occurring, it alerts the security administrator by generating an alarm. Also, IDS can initiate a proper response to the malicious activity.

Intrusion detection can be categorized into two methods: anomaly detection and misuse detection. Anomaly detection is the method of monitoring the network for deviations from normal behavior while misuse detection (also called signature-based detection) uses databases that contain signatures or patterns of known attacks [11].

Huang and Lee [12] proposed an intrusion detection system against several types of attacks in MANETs. Their paper is based on their previous work on anomaly detection which used cross-feature analysis to detect intrusions in a MANET [13]. Their latter work can be divided into two approaches: one based on detecting anomalies by implementing IDS on every node, and the other based on anomaly detection by implementing IDS for a cluster-based system.

In the first approach, they have used feature selection to identify anomalies. They have used a total of 141 features, and this approach can be used to detect new and unknown attacks. They have used certain features based on Monitoring node and Monitored node to classify and detect attack types. Then they further refined their work by proposing identification rules for identifying some well-known attacks such as black hole, random packet dropping, etc.

In the second approach, they proposed a cluster-based IDS as opposed to local IDS running on all the nodes to deal with limited power issue of MANET. Since running IDS on each node consumes battery power, the task of collecting network information is assigned to a single node in each cluster, which acts as the cluster head. Each cluster has a cluster head, called a Monitoring node, which monitors the other nodes in the cluster, which are called Monitored nodes. The cluster head can overhear the traffic from its neighbors by using the promiscuous mode in MANET routing algorithms. They also devised an election mechanism to select the cluster heads fairly in such a way that each node has an equal chance of being selected as the cluster head.

Results: Comparing the two approaches, it can be seen that the cluster-based IDS approach performs much better in terms of CPU speed up and network overhead as compared to the first approach of running IDS every node. Although the accuracy in terms of detection is minimally better in the first approach, the overall benefits of the second approach outweigh the first.

Trivedi et al. [14] proposed a reputation-based mechanism to deal with intrusion in MANETs. They have named this mechanism as RISM (reputation-based intrusion detection system for mobile ad hoc networks) and it is a modification of the CONFIDANT protocol [15]. RISM has been designed as a "semi-distributed nature" which implies that it is neither restricted locally nor immediately propagated to the whole network.

RISM has the following modules:

(a) **Monitor**: which takes the responsibility of monitoring the network. It collects network information at fixed time intervals, called as Timing Windows.
(b) **Reputation System**: which assigns reputation values to the nodes. The reputation system can assign a node to be either Normal, Suspicious, or Malicious. The reputation is assigned on the basis of a threshold for dropped packets,

called MaliciousDropThreshold. This MaliciousDropThreshold is flexible in the sense that its value can be updated according to the network traffic in each Timing Window.

(c) **Path Manager**: which calculates a new path when a node is deemed as Malicious.

(d) **Redemption and Fading**: which is a mechanism by which a node deemed as Malicious is given the chance to improve its reputation. This is implemented by carrying out a knock test to see if a Malicious reputed node behaves normally on receiving the knock test. If a Malicious node successfully passes the knock test, it can be moved to Suspicious category.

Results: RISM performs better than normal DSR in terms of packet delivery ratio up to a certain extent but when the number of malicious nodes increases, RISM incurs a routing overhead as compared to DSR because of recalculation of a new node when a malicious node is detected.

Nadeem and Howarth [16] proposed intrusion detection and adaptive response (IDAR), an IDS mechanism that deploys both anomaly detection and knowledge-based intrusion detection. This is an enhancement over their previous work in which they dealt with intrusion detection in a predetermined way. Their proposed mechanism implements an adaptive intrusion response after the intrusion has been detected. This adaptive response takes into account parameters such as attack severity, network degradation, and impact of the response action on the network performance.

IDAR employs a cluster-based IDS in which all nodes can be either manager node (MN), cluster head (CH), or cluster node (CN). IDAR uses two matrices for keeping track of the network. They are network characteristic matrix (NCM) and performance matrix (PM).

The architecture of IDAR consists of the following stages:

(a) **Network Monitoring and Data Collection**: In this phase, the CHs collect data from CNs and store them in NCM and PM.

(b) **Training**: In this phase, CHs continuously gather NCM and PM information and report to MN at fixed time intervals.

(c) **Testing**: Testing is carried out in four further phases as follows:

- **Intrusion detection**: MN uses anomaly-based intrusion detection to identify if any intrusion has occurred.
- **Attack identification**: This phase uses a rule-based approach to identify the attack. This is done with the help of a knowledge base maintained by IDAR.
- **Intruder identification**: In this phase, MN applies intruder identification rules that a specific for a known attack.
- **Adaptive intrusion response**: It consists of three actions: Isolation, Route around attacker, and No punishment.

Results: IDAR performs better when compared to fixed intrusion detection response as the overall network degradation of IDAR is lower as compared to fixed intrusion detection response. For severe attacks such as black hole attack and sleep deprivation, IDAR can isolate the attacker node for most of the time. For rushing attacks, choosing No punishment response by IDAR gives the most optimal network performance.

3.2 Authentication-Based Approach

Hu et al. [17] proposed a generic route discovery mechanism for handling rushing attacks. Rushing attacks are DOS attacks which prevent on-demand routing protocols to find routes longer than 2 hops. In rushing attacks, the attacker forwards Route Requests much faster than other nodes. This is possible because the attacker ignores delays at the MAC or the delays imposed by the routing protocol. Although one solution is to ignore delays at all nodes altogether, it can cause degradation in network performance because of the resulting collisions in the network. To tackle this, Hu et al. have proposed a Secure Route Discovery mechanism for defending against rushing attacks.

The proposed mechanism consists of three phases:

(a) **Secure Neighbor Detection**: This phase uses a three round mutual authentication protocol to determine if two nodes are neighbors so that they can communicate. This is carried out by deploying three messages:

- Neighbor Solicitation packet sent by the initiating node to a neighbor.
- Neighbor Reply packet sent by the neighbor on receipt of the previous packet.
- Neighbor Verification packet sent by the initiator which includes broadcast authentication of a timestamp and the link from source to the destination.

The protocol uses nonces to ensure freshness of the reply messages.

(b) **Secure Route Delegation**: In this phase, all nodes verify that secure neighbor detection protocols were executed correctly. A node receiving a Route Request verifies that the request came from its neighbor.

(c) **Randomized Message Forwarding**: In this phase, a node first collects a number of Route Requests and selects a random request to forward. This random selection is done to ensure that attacker cannot dominate all the routes returned.

Results: The proposed mechanism is able to detect alternate routes in case of a rushing attack most of the time as compared to existing on-demand routing protocols which are in general unable to deliver packets over paths longer than two hops. However, the proposed mechanism has very low packet delivery ratio as compared to DSR in normal traffic conditions. The packet overhead is also large in the proposed mechanism as compared to DSR.

3.3 Software Agent-Based Approach

Prathapani et al. [18] proposed the use of mobile honeypot agents to detect black hole attacks in Wireless Mesh Networks (WMNs). Honeypot agents are software agents that are used in IDS to detect malicious attackers. They are used to monitor the network and also can be used as decoys that lure attackers. Honeypots are deployed as mobile software agents that can traverse the entire network, and as such, they are not confined to individual nodes.

The use of honeypots in determining whether a node is malicious or not is illustrated as follows:

(a) A honeypot first places itself next to a node to be tested, called as testee node.
(b) It generates a Route Request bearing the address of a known destination node to the testee. That is, the honeypot already knows route the destination and it is trying to verify whether the testee node behaves normally or not.
(c) The testee node then sends its Route Reply in response to the Route Request from the honeypot.
(d) The honeypot node, in turn, sends a dummy data packet to the testee node to be sent to the known destination.
(e) Then, the honeypot queries the known destination whether it has received the dummy data packet via the testee.

Results: Simulations were carried out in AODV protocol using two kinds of topologies: Grid topology and Random topology. It is observed that employing the honeypot scheme increases network throughput significantly in Grid topology and in Random topology as compared to normal AODV under black hole attack.

4 Summary

MANETs face various challenges because of their inherent characteristics, their areas of deployment, and the limited resources. We have seen that attacks in MANETs try to exploit these native network properties and routing protocol deficiencies. The different kinds of attacks discussed in this paper can be summed up in Table 1.

We have also seen various approaches to counter the above-mentioned threats. A brief summary of the techniques discussed for handling the security threats in MANETs is shown in Table 2.

Table 1 Summary of attacks in MANET

Attack	Type of attack	Layer of attack	Security feature compromised	Effect
Eavesdropping	Passive	Physical layer	Confidentiality	Message interception
Black hole	Active	Network layer	Availability, integrity	Packet drop
Gray hole	Active	Network layer	Availability	Packet drop
Wormhole	Active	Network layer	Availability, integrity	Route manipulation
Rushing	Active	Network layer	Availability	Route manipulation
Sleep deprivation	Active	Network layer	Availability	Battery consumption

Table 2 Summary of security approaches

Authors	Mechanism	Routing protocol	Type of attack(s)	Effect
Huang and Lee [12]	IDS	AODV	Black hole, sleep deprivation	CPU speed up and low overhead
Trivedi et al. [14]	IDS	DSR	Packet drop	Improved packet delivery ratio
Nadeem and Howarth [16]	IDS	AODV	Black hole, gray hole, sleep deprivation, rushing	Successfully isolate attacker node
Hu et al. [17]	Authentication	DSR	Rushing	Detects alternate routes in case of rushing attack
Prathapani et al. [18]	Honeypot agents	AODV	Black hole	Improved throughput

5 Conclusion

The characteristic properties of MANETs (viz. trust-based relationship, lack of central coordination) make them vulnerable to different kinds of attacks. Moreover, MANETs have to often operate in challenging environments with limited resources (e.g., bandwidth, battery life). Hence, security is of prime importance in MANETs.

In this paper, we have looked at some of the attacks that can take place in MANETs and a few approaches to tackle these attacks. It is seen that more often than not, deploying security measures against these attacks acts as a double-edged sword in that the implementation cost of security mechanisms causes a compromise in overhead and/or efficiency of the network.

Still, research has been going on to optimize the cost of implementing these security measures [19, 20]. One future scope in this direction may be the application of Big Data to monitor, analyze, make inferences, and take decisions to tackle different attacks in MANETs.

References

1. Deng Hongmei, Li Wei, Agrawal Dharma P (2002) Routing security in wireless ad hoc networks. IEEE Commun Mag 40(10):70–75
2. Chandra P (2011) Bulletproof wireless security: GSM, UMTS, 802.11, and ad hoc securit. Elsevier, Armsterdam
3. Murthy CSR, Manoj BS (2004) Ad hoc wireless networks: architectures and protocols, portable documents. Pearson Education, London
4. Zhang Yongguang, Lee Wenke, Huang Yi-An (2003) Intrusion detection techniques for mobile wireless networks. Wirel Netw 9(5):545–556
5. Li Wenjia, Parker James, Joshi Anupam (2012) Security through collaboration and trust in MANETs. Mob Netw Appl 17(3):342–352
6. Yang H, Luo H, Ye F, Lu S, Zhang L (2004) Security in mobile ad hoc networks: challenges and solutions. IEEE Wirel Commun 11(1):38–47
7. Wu B, Chen J, Wu J, Cardei M (2007) A survey on attacks and countermeasures in mobile ad hoc networks. Wireless Network Security. Springer, US, pp 103–135
8. Kannhavong B, Nakayama H, Nemoto Y, Kato N, Jamalipour A (2007) A survey of routing attacks in mobile ad hoc networks. IEEE Wirel Commun 14(5):85–91
9. Hu Yih-Chun, Perrig Adrian, Johnson David B (2006) Wormhole attacks in wireless networks. IEEE J Sel Areas Commun 24(2):370–380
10. Anantvalee T, Jie W (2007) A survey on intrusion detection in mobile ad hoc networks. Wireless Network Security. Springer, US, pp 159–180
11. Nishani L, Biba M (2016) Machine learning for intrusion detection in MANET: a state-of-the-art survey. J Intell Inf Syst 46(2):391–407
12. Huang Y, Lee W (2003) A cooperative intrusion detection system for ad hoc networks. In: Proceedings of the 1st ACM workshop on security of ad hoc and sensor networks, ACM, New York
13. Huang YA, Fan W, LeeW, Yu PS (2003, May) Cross-feature analysis for detecting ad-hoc routing anomalies. In: Proceedings of the 23rd international conference on distributed computing systems, IEEE, pp 478–487
14. Trivedi AK, Kapoor R, Arora R, Sanyal S, Sanyal S (2013) RISM–reputation based intrusion detection system for mobile ad hoc networks. arXiv preprint arXiv:1307.7833
15. Buchegger S, Le Boudec JY (2002, June) Performance analysis of the CONFIDANT protocol. In: Proceedings of the 3rd ACM international symposium on mobile ad hoc networking and computing, ACM, pp 226–236
16. Nadeem A, Howarth MP (2014) An intrusion detection and adaptive response mechanism for MANETs. Ad Hoc Netw 13:368–380
17. Hu YC, Perrig A, Johnson DB (2003, September) Rushing attacks and defense in wireless ad hoc network routing protocols. In: Proceedings of the 2nd ACM workshop on wireless security, ACM, pp 30–40
18. Prathapani A, Santhanam L, Agrawal DP (2013) Detection of blackhole attack in a Wireless Mesh Network using intelligent honeypot agents. J Supercomput 64(3):777–804
19. Mitrokotsa A, Dimitrakakis C (2013) Intrusion detection in MANET using classification algorithms: the effects of cost and model selection. Ad Hoc Netw 11(1):226–237
20. Wang SH, Tseng CH, Levitt K, Bishop M (2007, September) Cost-sensitive intrusion responses for mobile ad hoc networks. In: International workshop on recent advances in intrusion detection, Springer, Berlin, Heidelberg, pp 127–145

A Survey: EMG Signal-Based Controller for Human–Computer Interaction

Tanuja Subba and Tejbanta Singh Chingtham

Abstract Human–computer interaction (HCI) has become one of the important aspects in human life. Signals generated from human body are biosignals and have huge potential to be used as an interface for human–computer devices. Multiple devices are present that recognize these biosignals which are generated during muscle contraction and converting those signals into some command to be used as an input to the HCI devices. However, the task can be acquired through biosignals which forms a neural linkage with the computer techniques like electroencephalogram (EEG), electrooculogram (EOG), and electromyogram (EMG). In past, there have been lots of studies wherein many researchers have used biosignals to control other device. EMG is hence one of the least explored mechanism form of biosignal to be deployed in HCI, and its studies are useful for neuromuscular system as certain diseases may slow down muscle contraction and muscle firing leading to paralysis of muscle.

Keywords EMG · HCI · Biosignals · Skeletal muscles · Neural linkage

1 Introduction

HCI is the one of the research areas that emerged in early 1980s, which has expanded rapidly, and it was previously known as a man–machine interaction. HCI focuses on the interface between user and the computer and deals with the design, execution, and assessment of computer system and other related systems that are for human use. Designing reciprocative interface to be helpful, valuable, resourceful, simple, and pleasant to use is important, in order to show an individual the benefits

T. Subba (✉) · T. S. Chingtham
Computer Science and Engineering Department,
Sikkim Manipal Institute of Technology, Majitar, Sikkim, India
e-mail: tanujasubba69@gmail.com

T. S. Chingtham
e-mail: chingtham@gmail.com

© Springer Nature Singapore Pte Ltd. 2019
H. K. D. Sarma et al. (eds.), *Advances in Communication,
Cloud, and Big Data*, Lecture Notes in Networks and Systems 31,
https://doi.org/10.1007/978-981-10-8911-4_13

of HCI devices [1]. The researchers observe the way human interacts with the computer system and design new technologies and interface that let human and computers to interact in novel ways [2]. Some of the examples of popular HCI techniques are image processing, speech recognition, biosignal processing, etc. HCI's goal is to minimize the differences between the human's goal of what they want to achieve and the understanding level of computer to perform the task. It relates knowledge from both the human and machine side. Due to its multidisciplinary nature, people with different study areas contribute to its success. Figure 1 shows the areas where HCI can be implemented with distinctive importance.

EMG is an electromedical procedure for estimating and capturing the biosignals produced during skeletal muscle contraction. This procedure is performed using electromyography, to produce an electrical record or signal called electromyogram [3]. Electromyography identifies the electrical signals when the brain activates the muscle cells voluntary or involuntary. The EMG signal has a unique gesture signature when it is captured with different movement and this gesture can act as an input for click point graphics and for many other applications [4]. The resulted signal can be classified and analyzed to check medical disabilities of human movement. The neurons of a human body transmit neurological and electrical signals that make muscle to contract, and an EMG translates these signals to graphs, sound, or numerical values that can be interpreted by analyst. EMG's signal can be easily acquired using electrodes, and it is of two types, dry electrode or a surface electrode which captures the signal when the electrodes are placed directly above

Fig. 1 Disciplines contribute to HCI [21]

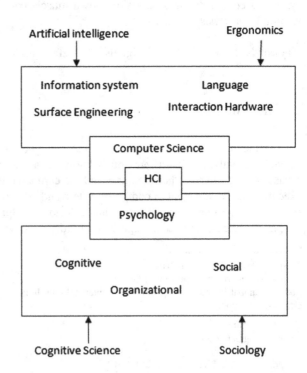

the muscle, the recording displays the difference between two placed electrodes and hence, requires more than one electrode. The second is gel or inserted EMG; these are the electrodes which are implanted inside the skin; therefore, for the interface between the skin and electrodes, electrolytic gel is used [5]. A needle electrode and fine wire electrode is the example of inserted electrode. Needle electrode is used in clinical areas, and the tip of the electrode is bare and used for the surface detection. Fine wire electrode is easily implanted in and withdrawn from the skeletal muscles, and is less painful then needle electrode. Thus, EMG which is related to muscle signal is very useful in terms of biomedical applications where it is possible to diagnose neuromuscular disease and many other disorders of motor control.

2 EMG Used for HCI

Studies are being carried out for the use of EMG signals in order to identify disabilities as there are many who are suffering from medical disabilities in terms of motor, neuron, muscle, etc. [6]. Therefore, EMG signals are not only used for identifying neuromuscular disorder but can also be used as a control signals for prosthetic devices [7]. It is the least explored compared to other biosignals like EEG, EOG. EMGs are considered to be a new and natural process of HCI as the induced signals from human muscle movement during its contraction represents neuromuscular movement that can be captured and filtered to command some specific devices. The applications of EMG signals are in medical area, human–computer interaction, etc. EMG can sense isometric muscular activity, i.e., when muscle fires but there is no joint movement which makes classification of subtle motionless gestures and control interfaces possible [8]. The EMG signal has different signatures, i.e., gesture can match but their characteristics like EMG signals are different in terms of age, muscle development, and sign. The one demerit of EMG signal is it contains a different type of unwanted signal called noise caused by external and internal environment and hence preprocessing is required to sort out the unwanted signal.

3 Related Works

Researchers have worked on regarding how EMG signal is used to command some other devices like prosthetic arm, robots, or to enable people with certain disabilities. These are shown in following paper.

In 1996, Koike and Kawato [9] developed an interface that models an arm and controls a robotic hand. The aim of this paper was to learn and enhance the motion capability. The technique used is Artificial Neural Network (ANN) that constructs forward dynamics model of the human arm. The result of the developed model was able to understand signal reading of EMG signals for instantaneous amount of movement.

In 2000, Alsayegh [10] proposed an EMG-based signal with the use of arm muscles, medial deltoid (MD), anterior deltoid (AD), biceps brachii (BB), which recognized 12 arm gestures. EMG signal processing is based on arm gestures having unique temporal coordination. The classification technique used is context-dependent classification with Bayes' theorem. Not only the unique arm gesture by using EMG signal was developed, there were various researchers working in the field of EMG for the people suffering with motor disabilities like hand paralysis, leg paralysis, etc.

In 2004, Kim et al. [11] proposed a natural means of human computer interaction induced by human arm's muscle movement. The captured EMG signal was used to computer command control. An online EMG MOUSE system was developed that controls cursor movement. The movement is interpreted for six different commands. The classification of the signal is done using fuzzy min–max neural network (FMMNN).

In 2005, Moon et al. [12] projected a wearable EMG device based on HCI for the wheelchair user. The developed device was able to serve people with C4 and C5 spinal cord injury. EMG signal was acquired by right, left, and both shoulder muscle motion. The wearable device directly generates mean average value (MAV) from raw EMG when shoulder muscle is contracted. Microcontroller is used to convert MAV signal to digital using analog-to-digital converter. Bluetooth module is used to send the matched result to device controller (wheelchair). The following year one more paper regarding people suffering from motor disabilities was presented.

In 2006, Kim et al. [13] developed an interface that relies on EMG signal acquired from human face during contraction of muscle. Electrodes are placed around forehead, cheeks, and eyes. The subject was made to perform some actions like blinking of eyes, clenching of teeth, wrinkling of forehead, and frowning. The signal is acquired and analyzed using linear prediction coefficient (LPC), and LPC entropy was calculated to find the characteristics information contained in the measured signal. For pattern recognition, hidden Markov model (HMM) is used. Same year some were working on hand gesture recognition.

In 2006, Naik et al. [14] proposed a method that identifies hand gestures with different electromyogram signals separated from electromyogram using independent component analysis (ICA). The paper explains and analyzes separation of the EMG signals by ICA for interpreting different hand gestures and actions.

After the recognition of hand gestures and enabling motor disabilities, in 2008 Kim et al. [15] proposed modification of a RC car. The control of a car was done by different user's hand signs. The features in the EMG signal were classified and analyzed to four different hand signs, and these signs were classified into four set of classes. The four classes were assigned to perform some commands for RC car. For feature extraction, RMS was used calculated by observing last 16 incoming values. For classification, KNN and Bayes' theorem were combined using decision tree and purpose a control the car via PC.

Similarly in 2009, Ren et al. [16] studied an electromyogram based on HCI. This paper showed a control system using forearm electromyography that is proposed for computer peripheral control and artificial prosthesis control. The system intends to realize the commands of six pre-defined hand poses, i.e., up, down, left, right, yes, and no. Power spectral density (PSD) is used to measure signal power intensity, and for classifier the Bayesian classifier is used for extracting feature. In the same year, Ahsan et al. [8] classified EMG signal techniques to improve interface for disabled people. This paper discusses various methodologies and techniques for interpreting EMG signal.

Researchers extended their study to multistep EMG classification; in 2010 Barreto et al. [17] proposed a system that intended to assist differently abled people. The experiment was performed for the people paralyzed from the neck down. Various interfaces like point and click graphics helped the subject to interact with computer and also communicate with other people. The EMG signal is generated using facial muscle for cursor movement.

In 2011, researchers studied surface EMG in depth and has found an excellent result; Ishii [18] studied about myoelectric prosthetic. The gestures acquired by arm or hand is identified and distinguished with the help of an EMG signal with the use of surface Electromyogram. For identification of motion, neural network is used.

In 2012, Tsujimura et al. [19] studied Hand Sign Classification. The EMG signal is acquired from forearm. The paper purposed to design a system that identifies finger motion to develop human–machine interface. Analysis of forearm EMG signal classifies and distinguishes the hand signs. It is also found that when fingers are moved, some of the muscles of forearm also work and generate EMG signal.

Researchers studied through multichannel surface EMG signals, and in 2014 Li et al. [3] showed HCI system based on the multichannel SEMG. In this paper, EMG signal is acquired using hand gestures and is classified for command control. This surface EMG signal controls quadcopter flight. The paper defines four different gestures to accurately achieve the real-time interfacing. The results showed have high accuracy for HCI using SEMG. Auto-regression method is used for analysis of SEMG signal, and the classification is done using back propagation technique.

In 2015, Mehaoua et al. [20] designed a system that controls multimedia player using EMG signal. The design was simple, and the working of the system was efficient and flexible. The paper aims to define an efficient and effective control system that can help and simplify people suffering from hand amputee by making them control device like media player using EMG signal acquired during con-traction of forearm. The electrical potential generated that is the different EMG signal generated during contraction allows controlling a media player like start, stop, and switching the video. Signal detection during muscle contraction is done it is rectified, filtered, and transformed into usable form. After this, the system was enhanced by adding commands like start, stop, previous, next, and pause.

4 Summary of Survey

The survey paper focuses on evaluation and detection of an EMG signal and use of this system for real time. There are many classification methodologies and techniques based on neural network to classify EMG signal. Some of the techniques are as follows.

4.1 Back Propagation Neural Network

This algorithm is used on multichannel SEMG [3] for hand gesture identification. The hand gestures control quad copter flight using extracted feature of EMG signal. Back propagation neural network uses neural nets to solve problem. It has three parts: building, training, and classification. Building is defined by input and output of the system, and training is defined when the weights are identified and modified.

4.2 Fuzzy Min–Max Neural Network

It is a classifier built using hyperbox fuzzy sets. The hyperbox fuzzy set contains n dimension pattern space and is completely defined by its minimum and maximum point. The hyperbox and the combination of minimum and maximum point define the fuzzy sets. This classifier is used by Kim et al. [11] for online EMG mouse to control computer cursor. The control is done using EMG signal acquired from arm's muscle. This signal was able to control the motion of mouse. The movement is interpreted as six pre-defined motions as per the actual mouse.

4.3 Hidden Markov Model

This is a statistical model for modeling generative states with hidden states. In hidden Markov model, a sequence of emission is observed but sequence of states is hidden and works with multiple probabilities. Ki-Hong et al. [13] used this model to develop an interface using EMG signal from human face. The stand-alone interface system was implemented, and the subjects (people as volunteers) were able to make the wheelchair turn left, right, forward, and backward by simple action provided by them.

Table 1 Summary of EMG classifiers

Classifier used	Description
Back propagation neural network [3]	• 93% success rate in the multichannel SEMG of the hand gesture recognition • Auto-regressive model method is used
Hidden Markov model [11]	• 97% success rate in developing an interface using human face • Subject was able to turn left, right, forward, and backward
Fuzzy min–max neural network [13]	• Six pre-defined motions of cursor were classified • 90% and above success rate for pattern recognition of each motion
Bayes' network [10, 15]	• Classification is done in probabilistic statistical model • Classification rate reported was 96%
	• KNN classifier and Bayes' classifier were combined for classification of EMG signal • 94% success rate for classification of hand gesture

4.4 Bayes' Network

Bayes' network is a probabilistic graphical model that represents a random variable in directed acyclic graph. Alsayegh [10] used this network to present an EMG-based interface for human and machine. The use of three muscles from arm, i.e., medial deltoid (MD), anterior deltoid (AD), and biceps brachii (BB) provides 12 different arm gestures by sensing the activities of the muscles. EMG signal processing is based on arm gestures having unique temporal coordination. The classification technique is defined by unique gesture's temporal signature used is context-dependent classification. Another researcher Kim et al. [15] developed similar interface using EMG signal for hand gesture by integrating the classifier K-nearest neighbor and Bayes' network.

Table 1 provides the summary of the survey in accordance with the methodologies used in various papers. It provides the description of the success rate resulted by the use of classification techniques.

5 Conclusion

Developing better human–computer interface will help improve quality of life of people suffering from physical disabilities. EMG signal is one of the natural techniques that captures electrical signals from human body for the use of HCI and provides an interface for human and computer to interact appropriately. This survey paper focuses on the work of various researchers; the methodologies are used for the classification of EMG signal. Therefore, it can be concluded from the survey of various papers that neural network has been used as a prominent classification technique of EMG signal for HCI.

For future works, newly enhanced classification techniques can be developed besides neural network, a work can be done in creating lightweight EMG signal.

References

1. Dix A (2009) Human-computer interaction, pp 1327–1331. Springer, US
2. Andurkar AG, Andurkar RG (2015) Human-computer interaction. Int Res J Eng Technol (IRJET) 2(6)
3. Li H, Chen X, Li P (2014) Human-computer interaction system design based on surface EMG signals. In: Proceedings of the 6th international conference on modelling identification & control (ICMIC), pp 94–98. IEEE, Dec 2014
4. Chowdhury RH, Reaz MB, Bakar AA, Hasan MS (2013) Muscle Technol 6(12):2192–2196
5. Day S (2002) Important factors in surface EMG measurement. Bortec Biomedical Ltd. Publishers, pp 1–17
6. Barreto AB, Scargle SD, Adjouadi M (2000) A practical EMG-based human-computer interface for users with motor disabilities. J Rehabil Res Dev 37(1):53
7. Ali AA, Albarahany A, Quan L (2012) EMG signals detection technique in voluntary muscle movement. In: 6th international conference on new trends in information science and service science and data mining (ISSDM). IEEE, 2012, pp 738–742, Oct 2012
8. Ahsan MR, Ibrahimy MI, Khalifa OO (2009) EMG signal classification for human computer interaction: a review. Eur J Sci Res 33(3):480–501
9. Koike Y, Kawato M (1996) Human interface using surface electromyography signals. Electron Commun Jpn (Part III: Fundam Electron Sci) 79(9):15–22
10. Alsayegh OA (2000) EMG-based signal processing system for interpreting arm gestures. In: 10th European Signal processing conference, pp 1–4. IEEE, Sept 2000
11. Kim JS, Jeong H, Son W (2004) A new means of HCI: EMG-mouse. In: IEEE International conference on systems, man and cybernetics, vol 1, pp 100–104. IEEE, Oct 2004
12. Moon I, Lee M, Chu J, Mun M (2005) Wearable EMG-based HCI for electric-powered wheelchair users with motor disabilities. In: ICRA 2005. In: Proceedings of the 2005 IEEE international conference on robotics and automation. IEEE, pp 2649–2654, Apr 2005
13. Kim KH, Yoo JK, Kim HK, Son W, Lee SY (2006) A practical biosignal-based human interface applicable to the assistive systems for people with motor impairment. IEICE Trans Inf Syst 89(10):2644–2652
14. Naik GR, Kumar DK, Singh VP, Palaniswami M (2006) Hand gestures for HCI using ICA of EMG. In: Proceedings of the HCSNet workshop on use of vision in human-computer interaction, vol 56, pp 67–72. Australian Computer Society, Inc., Nov 2006
15. Kim J, Mastnik S, André E (2008) EMG-based hand gesture recognition for realtime biosignal interfacing. In: Proceedings of the 13th international conference on intelligent user interfaces, pp 30–39. ACM, Jan 2008
16. Ren JR, Liu TJ, Huang Y, Yao DZ (2009) A study of electromyogram based on human-computer interface. J Electron Sci Technol China 7(1):69–73
17. Ren P, Barreto A, Adjouadi M (2010) Multi-step EMG classification algorithm for human-computer interaction. In: Innovations in computing sciences and software engineering. Springer, Netherlands, pp 183–188
18. Ishii C (2011) Recognition of finger motions for myoelectric prosthetic hand via surface EMG. INTECH Open Access Publisher
19. Tsujimura T, Yamamoto S, Izumi K (2012) Hand sign classification employing myoelectric signals of forearm. Curr Appl Future Chall 309

20. Hammi MT, Salem O, Mehaoua A (2015) An EMG-based human-machine interface to control multimedia player. In: 17th international conference on e-health networking, application & services (HealthCom), pp 274–279. IEE, Oct 2015
21. Human computer interaction: an overview. http://www.ee.cityu.edu.hk/~hcso/ee4213_ch1.pdf

IEEE 802.15.4 as the MAC Protocol for Internet of Things (IoT) Applications for Achieving QoS and Energy Efficiency

Dushyanta Dutta

Abstract Medium Access Control (MAC) protocol will play a very critical role in achieving the desired Quality of Service (QoS). The other crucial factor a MAC control is the total amount of energy spent by the network, which is very essential in determining the lifetime of the network. This paper argues that IEEE 802.15.4 can act as a suitable MAC protocol for application such as Internet of Things. The paper presents an overview on IEEE 802.15.4 protocol's working mechanism. Further, the paper also discusses certain issues need to be overcome for successful deployment of IEEE 802.15.4 in Internet of Things (IoT) to get the desired performance.

Keywords IEEE 802.15.4 · Quality of service (QoS) · Internet of Things (IoT)

1 Introduction

The rapid growth in Internet technology and tiny electronic devices in today's world has lead to emergence of a new technology term as Internet of Things (IoT). The goal of Internet of Things is to connect globally all electronic devices that are capable to communicate. This technology chooses Internet as a medium of communication so as the user can access its devices remotely from any Web-based application. The technology also allows these devices to send its data gathered from its surrounding to its user sitting remotely [1–4].

The Medium Access Control (MAC) is an integral part of communication system and will play a vital role in successful deployment of such network. The devices associated in Internet of Things are mostly small devices such as sensor devices. These devices generate low data rate and have radio transceivers to communicate with battery as its source of energy [5, 6].

IEEE 802.15.4 protocol which defines the Physical and MAC layer was designed for devices such as which generate low data rate and consume low power

D. Dutta (✉)
Kaziranga University, Jorhat, Assam, India
e-mail: dushyanta@kazirangauniversity.in; minkutezu@gmail.com

© Springer Nature Singapore Pte Ltd. 2019
H. K. D. Sarma et al. (eds.), *Advances in Communication,
Cloud, and Big Data*, Lecture Notes in Networks and Systems 31,
https://doi.org/10.1007/978-981-10-8911-4_14

127

[7, 8]. Therefore, we give an argument that IEEE 802.15.4 can be an ideal MAC protocol for application such as Internet of Things.

The paper in Sect. 1 gives an introduction on Internet of Things; Sect. 2 gives an overview on IEEE 802.15.4 protocol. Section 3 discusses certain challenges need to be addressed for successful deployment of IEEE 802.15.4 in Internet of Things and finally Sect. 4 draws the Conclusion of the paper.

2 Overview on IEEE 802.15.4

IEEE 802.15.4 PHY/MAC is designed for network such as Personal Area Networks (PANs). Nodes in this network operate in synchronized way by synchronizing itself to a node called coordinator. The protocol can operate for three network topologies —star, cluster tree, and mesh topology.

IEEE 802.15.4 MAC performs its channel access mechanism either in beacon-enabled mode or non-beacon-enabled mode. The beacon-enabled mode uses the slotted Carrier Sense Multiple Access with Collision Avoidance (CSMA/CA) while in non-beacon enable the unslotted CSMA/CA is used. An optional energy-saving mechanism can be implemented in the beacon-enabled mode using the concept of Duty Cycle. Duty Cycle mechanism allows the transceivers in the nodes to be either in Active or Sleep Modes in a synchronized way.

The transceivers remain active during the Superframe whose boundary is bounded by beacon frame, generated by the coordinator at regular interval. The size of the beacon frame is the interval between two consecutive beacons called Beacon Interval (BI) and $BI = 15.36 * 2^{BO}$ ms where $0 \leq BO \leq 14$. The size of the active period is called Superframe Duration (SD) and $SD = 15.36 * 2^{SO}$ ms, where $0 \leq SO \leq 14$ (Fig. 1).

All operations in the slotted CSMA/CA algorithm are aligned to discrete time slots with duration of 0.32 ms which also corresponds to a backoff time slot. When a packet is generated, the slotted CSMA/CA algorithm becomes operational and performs the following steps:

1. Initialize the variables—*contention window (CW)*, the **number of backoff stages (NB)**, and the **backoff exponent (BE)** as $CW = 2$, $NB = 0$ and $BE = macMinBE$ where the default *macMinBE* = 3.
2. Before sensing the channel, a node waits for a random number of backoff time slots uniformly distributed in the range $[0, (2^{BE} - 1)]$.
3. Sense the wireless medium for a **Clear Channel Assessment (CCA)**.
4. If the medium is found busy, set $NB = NB + 1$, $BE = BE + 1$ (BE can be incremented up to **macMaxBE** = 5) and $CW = 2$. If the number of backoff stages *NB*, exceeds the maximum allowed value **macMaxCSMABackoff** (with default value 4) drop the frame and exit. Otherwise, return to **Step 2**.
5. If the medium is found free and $CW = 2$, then set $CW = CW - 1$ and return to **Step 3**. If the medium is found free and $CW = 1$, then set $CW = 0$ and transmit the frame.

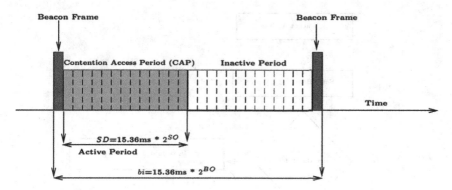

Fig. 1 An example of the superframe structure [7]

The CSMA/CA algorithm supports both retransmission and non-retransmission of the data frames. The retransmission scheme is based on acknowledgements. When retransmissions scheme is enabled, the destination node sends an acknowledgement immediately after receiving a data frame. The frames which are not acknowledged have to be retransmitted. The maximum allowed retransmission is up to *aMaxFrameRetries*, which has a default value of **3**, before a frame is dropped. For retransmitting a frame, the CSMA/CA algorithm starts from **Step 1** (Fig. 2).

3 Challenges on Successful Deployment of IEEE 802.15.4

In this section, some of the challenges are discussed that will arise during the deployment of IEEE 802.15.4 devices in Internet of Things.

1. **Identification of appropriate Duty Cycle for IEEE 802.15.4 devices to deploy in Internet of Things**.

   Devices connected to IoT will have limited source of energy. The life of a network depends on the rate of energy consumed by devices. Therefore to sustain the network's life, the consumption of energy is needed to be done judiciously. The use of low duty cycle in IEEE 802.15.4 is a promising mechanism for extending the life of the network. However, with the use of low duty cycle, the network performance may degrade by throughput and latency and may not fulfill the QoS requirement of the application [9, 10]. The appropriate duty cycle has to be chosen based on the prevailing traffic load. The traffic generated by IoT devices will be of non-homogeneous nature.

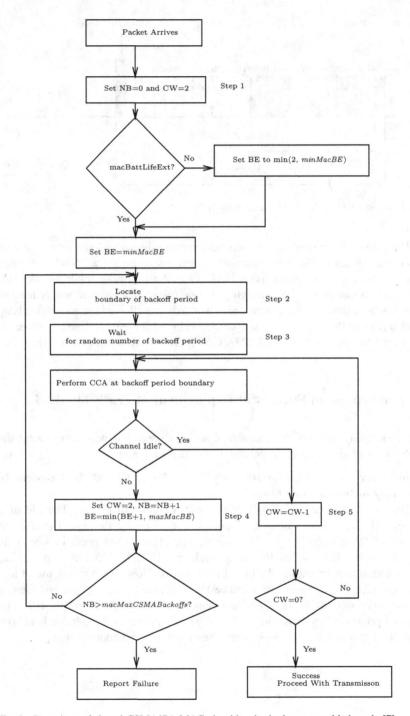

Fig. 2 Operations of slotted CSMA/CA MAC algorithm in the beacon-enabled mode [7]

2. **Synchronization of IEEE 802.15.4 devices appropriately for Multihop Deployment of devices in Internet of Things**.
To operate with duty cycle mechanism, IEEE 802.15.4 devices required to synchronize itself to a central device acting as the coordinator device. These devices synchronize with beacon frame transmitted at regular interval by the coordinator. Since these devices will have a very limited radio range to communicate, in multihop scenario there will be several coordinator devices. To avoid beacon frame collision, an efficient synchronization scheme is required for successful deployment of IEEE 802.15.4 devices in multihop scenario in Internet of Things. Therefore, an appropriate synchronization scheme is required to avoid beacon frame collision.

3. **Identification of appropriate MAC parameters value for successful deployment of IEEE 802.15.4 devices in Internet of Things**.
IEEE 802.15.4 MAC has number of parameter associated with the CSMA/CA protocol used for channel access. Value of these MAC parameters can determine the network performance by Energy Efficiency and QoS. Therefore, determining proper value for this parameter is essential to get the desired performance.

4 Conclusion

This paper highlighted that IEEE 802.5.4 protocol can be utilized as a promising MAC protocol for Internet of Things. It gives argument on the ground that device connected to Internet of Things will mostly be of tiny devices generating low data rate and IEEE 802.15.4 was designed for application consisting such devices. Secondly, the paper gives an overview in IEEE 802.15.4 and also discussed certain challenges need to be overcome for successful deployment of IEEE 802.15.4 devices in Internet of Things.

References

1. Gubbi J, Buyya R, Marusic S, Palaniswami M (2013) Internet of things (IoT): a vision, architectural elements, and future directions. Future Gener Comput Syst 29(7):1645–1660
2. Kelly SDT, Suryadevara NK, Mukhopadhyay SC (2013) Towards the implementation of IoT for environmental condition monitoring in homes. IEEE Sens J 13(10):3846–3853
3. Ho C, Shen TY, Hsu PY, Chang SC, Wen SY, Lin MH, Wu CM (2016) Random soft error suppression by stoichiometric engineering: CMOS compatible and reliable 1 Mb HfO$_2$-ReRAM with 2 extra masks for embedded IoT systems. In: IEEE symposium on VLSI technology, pp 1–2. IEEE
4. Lazarescu MT (2013) Design of a WSN platform for long-term environmental monitoring for IoT applications. IEEE J Emerg Sel Top Circuits Syst 3(1):45–54
5. Suriyachai P, Roedig U, Scott A (2012) A survey of MAC protocols for mission-critical applications in wireless sensor networks. IEEE Commun Surv Tutorials 14(2):240–264

6. Rajandekar A, Sikdar B (2015) A survey of MAC layer issues and protocols for machine-to-machine communications. IEEE Internet Things J 2(2):175–186
7. IEEE TG 15.4 (2003) Part 15.4: wireless medium access control (MAC) and physical layer (PHY) specifications for low-rate wireless personal area networks (WPANs). IEEE Std., New York
8. Ramachandran I, Das AK, Roy S (2007) Analysis of the contention access period of IEEE 802.15. 4 MAC. ACM Trans Sensor Netw (TOSN) 3(1):4
9. Dutta D, Karmakar A, Saikia DK (2014) Determining duty cycle and beacon interval with energy efficiency and QoS for low traffic ieee 802.15.4/ZigBee wireless sensor networks. In: Advanced computing, networking and informatics, vol 2. Springer International Publishing, pp 75–84
10. Dutta D, Karmakar A, Saikia DK (2014) An analytical model for IEEE 802. 15.4/ZigBee wireless sensor networks with duty cycle mechanism for performance prediction and configuration of MAC parameters to achieve QoS and energy efficiency. Int J Comput Appl 102(5)

HMM-Based Speaker Gender Recognition for Bodo Language

Chandralika Chakraborty and Pran Hari Talukdar

Abstract Speech, the act of speaking, is the most natural way of exchanging information between homo sapiens. Speech primarily conveys the message via words, spoken by the speaker. Speech also conveys the emotion with which the speaker speaks, speaker's health condition, gender of the speaker, and also the language in which the speaker is speaking. Systems which aim to recognize the speaker-related information in speech signals through an extraction and characterization process are called speaker recognition systems. Speaker recognition applications are becoming common and useful nowadays as many of the modern devices are designed and produced for the convenience of the general public. Speaker recognition systems are developed for many indigenous languages. Application of hidden Markov models (HMMs) to speaker recognition has seen considerable success and gained much popularity. This paper presents an attempt made toward developing a speaker gender recognition system. A model built using Hidden Markov Model Toolkit (HTK 3.4.1) has been trained and tested on sample speech of either gender in Bodo language, and results show good recognition.

Keywords Speaker gender recognition · Bodo · Hidden Markov model
Acoustical analysis · Mel frequency cepstral coefficient · Hidden Markov model
training · Recognition testing · Performance evaluation

C. Chakraborty (✉)
Department of Information Technology, Sikkim Manipal Institute of Technology,
Sikkim Manipal University, Majitar, Sikkim, India
e-mail: chandralika.c@gmail.com

P. H. Talukdar
Kaziranga University, Koraikhuwa, Assam, India
e-mail: phtassam@gmail.com

© Springer Nature Singapore Pte Ltd. 2019   133
H. K. D. Sarma et al. (eds.), *Advances in Communication,*
Cloud, and Big Data, Lecture Notes in Networks and Systems 31,
https://doi.org/10.1007/978-981-10-8911-4_15

1 Introduction

Speech or verbal communication among humans is arguably the reflection of the highest order of intelligence. Nobody knows for certain when or how speech communication started. Over the years, through evolution and associated changes in physiology, climate, and society, the current level of intelligence transfer has been achieved. An innately intelligent activity, speech communication, however, is not a monopoly of humans only, but other animals like dolphins and elephants are also known to communicate with sound. However, no other forms of life are known to have speech-based communication system as developed as man.

Speech signal is produced through semantic, linguistic, articulatory, and acoustic transformations. It is the properties of these transformations which are identified as the acoustical properties of the speech signal. Further, the anatomical differences in the vocal tract and individual speaking habits together with these acoustical properties are used if we want to find out who is speaking. A speaker recognition system takes into account these differences and discriminates between speakers [1], say as male speaker or female speaker. *Speaker recognition* is the identification of the person who is speaking from the characteristics of the voice sample of the speaker.

The general area of speaker recognition comprises two fundamental tasks—speaker identification and speaker verification. *Speaker identification* is the task of identifying the speaker from a set of known speakers. *Speaker verification* is the task of verifying the claimed identity of the speaker, and such systems return a binary value in the form of a yes/no decision. Speaker recognition tasks are further distinguished based on the text of the speech used in the system. In a *text-dependent* system, the same word or phrase is used to train and test the system; every time the system is trained or tested. In a *text-independent* system, the training and testing speech is unconstrained.

The proposed work is an attempt at building a simple speaker recognition system using Hidden Markov Toolkit (HTK 3.4.1) [2] for Bodo language to help achieve text-independent speaker identification. Sect. 2 gives an overview of the previous work done in speaker recognition, Sect. 3 introduces the Bodo language. Section 4 defines the problem for developing the recognition system. Section 5 introduces the hidden Markov model and the Hidden Markov Toolkit 3.4.1. Section 6 explains the methodology for designing the speaker gender recognition system. Section 7 provides the findings of the work done, and finally, Sect. 8 gives the conclusion.

2 Previous Work

The features and the techniques that are applied to speech recognition for extraction and recognition of information from speech signal have also been used for speaker recognition. A major advance was the application of the hidden Markov model

(HMM)-based approach in 1980s, and it contributed largely to the acceptance of results and product development which are now being used by the telephone industry for voice dialer, voice response systems at banks, etc. These speech recognition systems are able to respond to non-specific speakers because of the technological advances that have made them possible. The success of the HMM-based approach to speech recognition and speaker recognition is evident from the fact that quite a few commercial applications have been built using this technique. Some of the most important and successful systems include *Sphinx*, a continuous-speech, and speaker-independent recognition system built using HMMs was developed by Kai-Fu Lee, which attained word accuracies of 71, 94, and 96% on a 997-word task [3]. An upgradation, *Sphinx-4*, a state-of-the-art speech recognition system created through joint collaboration between Carnegie Mellon University and others, is also popular [4]. *Julius*, an open-source large-vocabulary speech recognition software, named after Gaius Julius Caesar is currently used for both academic research and industrial applications.

Various speaker recognition systems have been developed for Indian languages like Assamese, Bengali, Hindi, Gujarati and Tamil. Deka et al. [5] proposed an approach for speech recognition using linear predictive cepstral coefficients (LPCC) and multilayer perceptron (MLP)-based artificial neural network with respect to Assamese. Patel and Virparia developed a speaker recognition system which accepts telephony commands in Gujarati [6]. The work makes use of a data set of 29 speaker-independent words for experimentation [6]. The implementation was done using HMM-based speech recognizer Sphinx4 toolkit. The system accuracy ranged from 72.41 to 96.55%. A connected Hindi digit recognition system using HTK was developed by Mishra et al. This work created a baseline recognizer, which provided encouraging results [7]. Vimala and Radha built a speaker-independent isolated speech recognition system for Tamil language [8].

However, the efforts directed at research are not concentrated on languages like Bodo, which had received very little attention.

3 Bodo Language

Bodo is a language which came out as a branch of its main stock known as Sino-Tibetan or Tibero-Chinese family of languages serially through its other sub-branches called Tibeto-Burman, Assam-Burmese, Bodo-Naga, and Bodo. It is used as language media in some of the districts of Assam, like Dhubri, Bongaigaon, Kokrajhar, to name a few.

Bodo phonemes consist of sixteen consonants and six vowels [9]. Bodo language was included in the eighth schedule of Indian constitution from 2005 and thus has recognition from central and state governments and their affiliated bodies for the development of its every aspect of the language and literature [10].

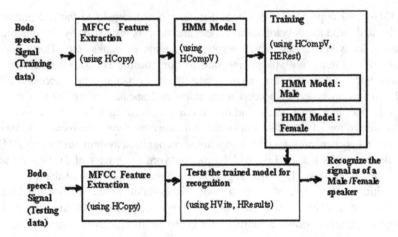

Fig. 1 Block diagram of the speaker gender recognition system for Bodo speech using HTK

4 Problem

The work reported here is aimed at developing a text-independent speaker recognition model (as shown in Fig. 1) capable of identifying the gender of the speaker. The problem can be formally stated as:

Given an arbitrary waveform, w_i, in a language L and its mel frequency cepstral coefficients (MFCC) set {m1, m2,.., m13}, to return a bi-valued classification MALE/FEMALE depending upon the gender of the speaker.

5 Hidden Markov Model

A Markov model depicts every observable event as a state. A hidden Markov model assumes the system to be modeled as a Markov process with unobserved states. Speech signals are normally continuous and are difficult and sometimes even unnecessary to determine how and when there is a transition from one abstract speech code to another. This transition from one speech code to another is depicted as a state transition. Hence, each state cannot be uniquely associated with an observable event. The outcomes or the observations are a probabilistic function of each state. The actual state sequence is not directly observable (that is hidden); it can only be approximated from the sequence of observations produced by the system. The HMM model used in this work is shown in Fig. 2.

Fig. 2 A HMM topology

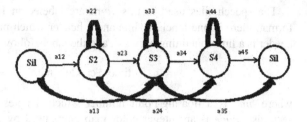

5.1 Hidden Markov Model Toolkit

The Hidden Markov Model Toolkit (HTK) is mainly used for speech recognition research, for building, training, testing hidden Markov models [2]. Facilities like result analysis are also provided by this toolkit [2]. The library modules and tools in HTK are available in C language. The toolkit also finds use in various applications like speech synthesis, character recognition, and DNA sequencing. HTK was originally developed at the Machine Intelligence Laboratory of the Cambridge University Engineering Department [2].

6 Methodology

The speech signals (of male as well as female speakers) are first recorded. Then, for each of the recorded speech files, a text is associated with its content. This is called labeling of speech files. In this work, wavsurfer is used for recording and labeling of speech files. The data set (comprising male and female speaker speech files) used here is recorded in a noise-free environment, labeled using wavsurfer and saved in . wav audio format. The data set consists of 100 samples each for male and female speaker voices. Eighty samples each of male and female speakers are taken for training the system, and the remaining for testing. All the samples (male as well as female) are of uniform utterances, ranging from 6 to 8 words. Each speech waveform is of 1 s duration. HSLab tool of HTK can also be used for recording and labeling. A sample waveform of a Bodo speech as generated in wavsurfer can be seen in Fig. 3.

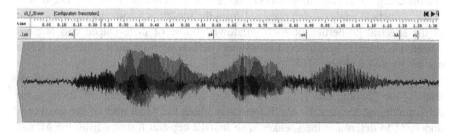

Fig. 3 A waveform

The speech files used in this work are labeled in ESPS label format. In this format, there is one label per line and a header which precedes the label data. The header is a line containing only a #, and the labels follow the header in the form [8]:

time   ccode   name

where *time* is a floating-point number which denotes the boundary location in seconds, *ccode* is an integer color map entry used by ESPS in drawing segment boundaries, and name is the *name* of the segment boundary. A sample ESPS format label file used for labeling all the speech samples in the data set is:

```
            #
      0.211651   125   si
      0.656355   125   bong
      0.929836   125   bong
      1.202128   125   thang
      1.272281   125   si
```

6.1 Acoustical Analysis

The toolkit, HTK, processes the speech waveforms after they are transformed into a sequence of feature vectors. This step is known as 'acoustical analysis.' The signal is first segmented in successive frames of a time frame 25 ms. Each frame is then multiplied by a windowing function, that is Hamming window. Hamming window helps to avoid discontinuities at the boundary of the window by shrinking the values toward zero. From each frame, features are extracted. One way of representing the features is called the mel frequency cepstral coefficient (*MFCC*). Human hearing is less sensitive to higher frequencies roughly above 1000 Hz [11]. Hence, considering this property of human hearing while extracting features can improve speech and speaker recognition. The model used in MFCC represents the frequencies onto the mel scale [12]. These parameters, such as the length of each frame and the kind of features to be extracted, must be specified in a configuration file. The configuration file is a text file which sets the parameters of the acoustical coefficient extraction. The original waveform is then converted to a set of acoustical vectors using the HCopy tool of HTK along with the configuration file. The output of the acoustical analysis step obtained from the toolkit is a compact representation of the spectral properties of each windowed frame of male-voiced and female-voiced waveform. The extracted acoustical coefficients from each windowed frame can be viewed with the HTK tool, HList.

A property of the cepstral coefficients is that the variance of different coefficients tends to be uncorrelated [11]. In a speaker recognition system, the focus is on determining the speaker. Hence, information about the vocal tract of the speaker is important to determine the speaker. The first 12 cepstral features from the MFCC extraction represent information about the vocal tract (the filter) which shapes the pulses of air produced by the glottis (the source). The shape of the vocal tract has particular filtering characteristics. The shape of a vocal tract is different for male speakers and for female speakers. Hence these 12 cepstral coefficients for each frame can be used to distinguish between a male and a female speaker. A thirteenth feature is also added which represents the energy in a frame. Energy correlates with phone identity. A speech signal varies from one frame to another frame. This change can be represented by adding a delta feature and an acceleration feature to each of the 13 features of each frame of the speech waveform [11]. The following coefficients are extracted for each signal frame:

- The first 12 MFCC coefficients and
- The null MFCC coefficient c0 which is proportional to the total energy in the frame.
- Thirteen delta coefficients, estimating the first-order derivative of [c0, c1,…, c12]
- Thirteen acceleration coefficients, estimating the second-order derivative of [c0, c1, …, c12].

So, this makes up a 39 coefficient vector for further processing.

This *39 coefficient vector* extracted from a female speech sample is as shown in Fig. 4.

6.2 Model Definition and Initialization

An HMM is used to model each of the acoustical events. Here, an acoustical event is a male-voiced waveform or a female-voiced waveform. For this, a topology for each HMM is first chosen. A five-state topology is adopted for each HMM. The model consists of three emitting states {S2, S3, and S4}; the first and the last states

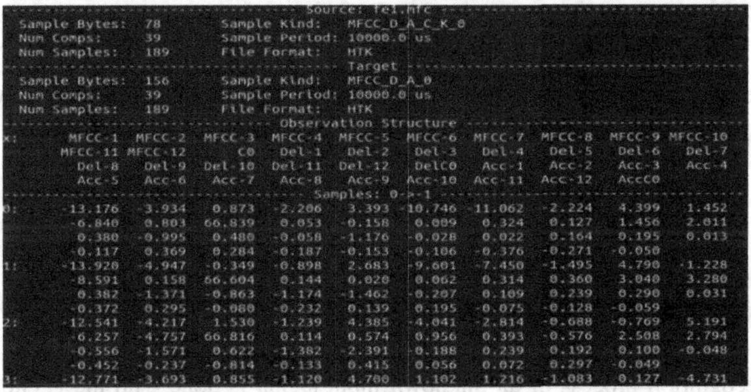

```
------------------------------------ Source: fe1.mfc ------------------------------------
 Sample Bytes:  78        Sample Kind:   MFCC_D_A_C_K_0
 Num Comps:     39        Sample Period: 10000.0 us
 Num Samples:   189       File Format:   HTK
------------------------------------ Target ------------------------------------
 Sample Bytes:  156       Sample Kind:   MFCC_D_A_0
 Num Comps:     39        Sample Period: 10000.0 us
 Num Samples:   189       File Format:   HTK
------------------------------------ Observation Structure ------------------------------------
x:     MFCC-1  MFCC-2  MFCC-3  MFCC-4  MFCC-5  MFCC-6  MFCC-7  MFCC-8  MFCC-9 MFCC-10
      MFCC-11 MFCC-12      C0   Del-1   Del-2   Del-3   Del-4   Del-5   Del-6   Del-7
        Del-8   Del-9  Del-10  Del-11  Del-12   DelC0   Acc-1   Acc-2   Acc-3   Acc-4
        Acc-5   Acc-6   Acc-7   Acc-8   Acc-9  Acc-10  Acc-11  Acc-12   AccC0
------------------------------------ Samples: 0->-1 ------------------------------------
0:    -13.176  -3.934   0.873  -2.206   3.393 -10.746 -11.062  -2.224   4.399   1.452
       -6.840   0.803  66.839   0.053  -0.158   0.009   0.324   0.127   1.456   2.011
        0.380  -0.995   0.480  -0.058  -1.176  -0.028   0.022   0.164   0.195   0.013
       -0.117   0.369   0.284  -0.187  -0.153  -0.106  -0.376  -0.271  -0.050
1:    -13.920  -4.947  -0.349  -0.898   2.683  -9.601  -7.450  -1.495   4.790  -1.228
       -8.591   0.158  66.604   0.144   0.020   0.062   0.314   0.360   3.040   3.280
        0.382  -1.371  -0.863  -1.174  -1.462  -0.207   0.109   0.239   0.290   0.031
       -0.372   0.295  -0.080  -0.232   0.139   0.195  -0.075  -0.128  -0.059
2:    -12.541  -4.217   1.530  -1.239   4.385  -4.041  -2.814  -0.088  -0.769   5.191
       -6.257  -4.757  66.816   0.114   0.574   0.956   0.393  -0.576   2.508   2.794
       -0.556  -1.571   0.622  -1.382  -2.391  -0.188   0.239   0.192   0.100  -0.048
       -0.452  -0.237  -0.814  -0.133   0.415   0.056   0.072   0.297  -0.049
3:    -12.771  -3.693   0.855  -1.120   4.700   1.102   1.216  -1.083   0.127  -4.731
```

Fig. 4 MFCC coefficients (of the speech sound ‘bong bong thang’ (female))

Fig. 5 HMM for a speech
waveform, 'za za ha'

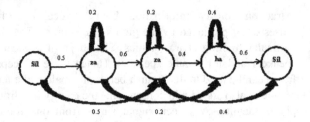

{S1 and S5} are non-emitting states, that is they have no observation function. The observation functions are single Gaussian distributions with diagonal matrices. The topology as shown in Fig. 2 is used for male speaker and also for female speaker waveforms. A sample HMM is shown in Fig. 5.

The HMM definition is described in HTK as a text description file. This HMM description file (prototype) is used by HCompV HTK tool for model initialization. In this proposed work, such a prototype has to be generated for a male-voiced and female-voiced waveform. The prototype for the two HMMs will be mentioned with headers ~h "ma" and ~h "fe".

6.3 HMM Training

Model re-estimation is performed with the HERest HTK tool, which estimates the optimal values for transition probabilities, mean, and variance vectors of each observation function. The re-estimation procedure is repeated for each HMM (one for male and another for female). In this procedure, the system is trained with various male speaker and female speaker voice samples. The training set contains 80 male and female voice samples each.

The basic architecture of the speaker gender identification system is to be defined through the task grammar. In HTK, the *task grammar* is a text file as given below:

```
$speaker = Female|Male;
($speaker)
```

The task grammar is compiled with the HParse tool, to obtain the task network. The speaker gender identification system is defined by this task grammar, the dictionary which contains the models to be recognized (male and female) and the HMM sets (the two HMM sets used in this system are named as: ma, fe).

6.4 Recognition Testing

The recognition process of an unknown input signal is done with the HVite tool, which matches the input observations (that is the acoustical observations stored as.

mfcc files) with the identification system's hidden Markov models. The testing set contains 20 male and female voice samples each. The system was tested for recognition using 10, 15, and 20 male and female voice samples each at different times.

6.5 Performance Evaluation

The performance of the system is evaluated with the HResults tool. The result of the identification is obtained in a file as male or female depending on the voice sample.

7 Findings

The speaker recognition system developed here is trained using a various number of speaker samples. The system is then tested for unknown inputs with a number of test samples. The output obtained in the different cases is represented in the form of a confusion matrix. Finally, the summary of the finding is provided in Table 5 which shows encouraging results.

Case I:

Training set = 40 voice samples each for male and female speaker
Testing = 10 voice samples each for male and female speaker (Table 1)

Case II:

Training set = 60 voice samples each for male and female speaker
Testing = 15 voice samples each for male and female speaker (Table 2)

Case III:

Training set = 80 voice samples each for male and female speaker
Testing = 20 voice samples each for male and female speaker (Table 3)

Case IV:

Training set = 60 voice samples each for male and female speaker
Testing = 20 voice samples each for male and female speaker (Table 4)

Table 1 Results of Case I

Actual	Predicted		
		M	F
	M	10	0
	F	0	10

Recognition = 100%

Table 2 Results of Case II

	Predicted	
Actual		M	F
	M	15	0
	F	0	15

Recognition = 100%

Table 3 Results of Case III

	Predicted		
Actual		M	F
	M	19	0
	F	0	20

Recognition = 97.5%

Table 4 Results of Case IV

	Predicted		
Actual		M	F
	M	19	0
	F	0	20

Recognition = 97.5%

Table 5 Summary of the results obtained

	Training data		Test data		Success count		Success rate
	Male	Female	Male	Female	Male	Female	
1.	40	40	10	10	10	10	**100**
2.	60	60	15	15	15	15	**100**
3.	80	80	20	20	19	20	**97.5**
4.	60	60	20	20	19	20	**97.5**

8 Conclusion

The work reported here is a speaker recognition system which recognizes the gender of the speaker using HTK for Bodo speech. The HMM model was built using five number of states. The speaker gender recognition system was trained and tested with proper samples, where each sample consisted of six to eight words, for performance evaluation. Acceptable results were obtained through detailed experimentation. It is observed that irrespective of the size of the training data set, increasing heterogeneity of the testing data degrades performance. However, performance matrix reveals success rates between 97 and 100%. Presently, work is on to develop a system with an eleven-state HMM, to recognize speaker's gender as male or female based on three-word and one-word samples.

References

1. Fu Z, Zhao R (2003) An overview of modeling technology of speaker recognition. In: IEEE proceedings of the international conference on neural networks and signal processing, vol 2, pp 887–891, Dec 2003
2. Young S et al (2009) The HTK book (for HTK version 3.4). Cambridge University Engineering Department
3. Lee K-F, Hon HW, Reddy R (1990) An overview of the SPHINX speech recognition system. In: IEEE transactions on acoustic speech and signal processing, vol 38, no 1
4. Admin. 'sphinx-4 application programmer's guide (2015). http://cmusphinx.sourceforge.net/sphinx4
5. Deka MK, Nath CK, Sarma SK, Talukdar PH (2011) An approach to noise robust speech recognition using LPC cepstral coefficient and MLP based artificial neural network with respect to Assamese and Bodo language. In: International symposium on devices MEMS, intelligent systems & communication (ISDMISC)
6. Patel J, Patel P, Virparia P (2014) Voice enabled telephony commands using Gujarati speech recognition. Int J Adv Res Comput Sci Softw Eng 3(12)
7. Mishra AN, Biswas A, Chandra M, Sharan SN (2011) Robust Hindi connected digits recognition. Int J Signal Process Image Process Pattern Recognit 4(2)
8. Vimala C, Radha V (2012) Speaker independent isolated speech recognition system for Tamil language using HMM. Procedia Eng 30:1097–1102
9. Boro MR (2008) The structure of Boro language. N.L. Publications, Panbazar, Guwahati
10. Boro MR (2007) The historical development of Boro language. N.L. Publications, Panbazar, Guwahati
11. Jurafsky D, Martin J (2014) Speech and language processing, 2nd edn. Pearson
12. Stevens SS, Volkmann J, Newman EB (1937) A scale for the measurement of the psychological magnitude pitch. J Acoust Soc Am 8(3):185–190. Bibcode:1937A SAJ....8..185S. https://doi.org/10.1121/1.1915893

Accurate Drainage Network Extraction from Satellite Imagery—A Survey

Ferdousi Khatun and Pratikshya Sharma

Abstract The extraction of the drainage hydrographical network is very important for various types of study such as hydrological analysis, geomorphology, environmental science, terrain analysis and still a research topic in the field of GIS. Drainage network is extracted through satellite image (e.g., digital elevation model) processing, contour map processing, and raster map processing. A raster map of an area contains many layers such as road network, building, forest area, waterbody, river pattern, text, and drainage pattern extracted from raster map is part of document image analysis. A toposheet or contour map contains the linear feature, namely elevation contour, waterbody, river network, text, and the extraction process of drainage line is time-consuming and traditional process. Due to the advances in satellite imagery, high-resolution digital elevation model (DEM) is captured by many satellites recently. The DEMs are advantageous over toposheet because it provides seamless provision of data with global coverage. Accurate drainage extraction from DEMs is used for morphometric analysis, hydrological analysis, terrain analysis, and many other areas in recent year across the world as DEM provides the fastest way to extract feature in various ways. This paper provides the evolution of satellite imagery and the accurate extraction of drainage network for various applications, namely geomorphometric analysis, hydrologic analysis, terrain analysis, and also describes the steps involved to extract drainage pattern from DEM, an up-to-date process.

Keywords DEM · ASTER-GDEM · SRTM · Cartosat-1 DEM

F. Khatun (✉) · P. Sharma (✉)
Computer Science and Engineering Department,
Sikkim Manipal Institute of Technology, Majitar, Sikkim, India
e-mail: ferdousi9@yahoo.com

P. Sharma
e-mail: pratikshya2007@yahoo.co.in

© Springer Nature Singapore Pte Ltd. 2019
H. K. D. Sarma et al. (eds.), *Advances in Communication,
Cloud, and Big Data*, Lecture Notes in Networks and Systems 31,
https://doi.org/10.1007/978-981-10-8911-4_16

145

1 Introduction

Water is the main and most important component on earth surface, and drainage network is the essential hydrologic, geomorphologic element for analysis. The patterns formed by the streams, rivers, and lakes in a particular drainage basin are known as drainage network or river system [1]. It has many application areas like hydrologic modeling of micro-watershed, groundwater prospect zone mapping, geomorphometric parameter analysis, water resource planning and management, flood hazard prediction and mitigation, river pattern change detection, irrigation management fields. But extraction of drainage pattern on flat surface and less complex terrain is still in research topic. The traditional process of generating the drainage map is from toposheet or contour map that depicts the large-scale detail of a geographic space. Generally, it contains five layers of information: river and waterbody as blue color, road as red, forest as green, contour as brown, and black color for text feature. However, channel network extraction from topomaps requires tedious time and cartography expert needed to provide subjective decision. In brief, the method to extract drainage from contour map is areal element removal, linear element extraction to produce linear feature map, thinning, dilation, color segmentation and generate segmented layer map [2]. But generation of separate layer map has a limitation; it will only work on digital contour map or historical map or high-quality toposheet because poor quality toposheet suffers from false color aliasing and mixed color problem. Also, available toposheet is old publication, 10–20 years back. River pattern also changes their position in some places due to landslide, flood, or many natural phenomena; thus, accurate drainage network is not provided by toposheet [3]. A new generation of digital elevation data is generated by remote sensing technologies due to advances in satellite technology, e.g., shuttle radar topographic mission (SRTM), interferometric synthetic aperture radar for elevation (IFSARE), advanced spaceborne thermal emission and reflection radiometer global digital elevation model (ASTER-GDEM V2), Cartosat-1 (Cartosat-DEM 1.0), synthetic aperture radar (SAR), so analysis became easier. Early satellite technology does not able to capture the DEM directly, so for hydrologic and geomorphologic analysis, DEM was generated from various satellites like QUICK BIRD, IKONOS, Landsat tm [4]. The most widely used satellite imagery ASTER-GDEM, SRTM-DEM, etc., is freely provided by USGS (http://earthexplorer.usgs.gov/) site with 30 m spatial resolution, and Cartosat-DEM is provided by ISRO (http://bhuvan.nrsc.gov.in/bhuvan_links.php) and opened the door for analyzing the data for various studies. The above-mentioned satellite has their specific configuration and is suitable for mountain area, flat surface, and medium complex terrain area for accurate river network extraction. The accuracy of SRTM–DEM and ASTER-GDEM is almost same [5]. In some cases, Cartosat-1 DEM sounds better than SRTM DEM for extracting drainage. DEMs with higher resolution provide detailed drainage networks that have greater impact on the drainage map analysis as statistical values become lower when resolution of DEM changes from fine to rude [6]. In this paper, we present an up-to-date overview of drainage network extraction process.

2 Evaluation of Satellite Imagery for Drainage Extraction

DEM plays a vital role for extracting drainage network. Before the year 2000, the DEMs were available at a global coverage in a 1 km resolution like GTOPO-30 (Global Topography in 30 arc sec). After that shuttle radar topography mission (SRTM, version 4, C-band DEM of 90 m resolution) and the advanced spaceborne thermal emission and reflection radiometer (ASTER, version 2, 30 m resolution) were launched that provided better resolution which solves the problem regarding spatial resolution. The hydrology analysis became more easier in INDIA after the launch of Cartosat-DEM (version 1, only for Indian territories) at 30 m in 2011. These data are freely available and easily downloaded from USGS and ISRO Web site. Various purchased stereo-images from Cartosat-1, landsat 7 ETM+, QuickBird, IKONOS, SPOT, and SAR are used for generating the DEM using software and can be applied for hydrologic analysis [4, 7–10]. Lot of studies are going on geomorphometric and hydrological analyses from DEMs. In India, most research is done best on DEM for river basin analysis, estimation of soil loss, water resource evaluation, and topographic characterization [11–14]. The high-resolution DEMs provide finer extraction of land surface component like drainage network, slope facets, and higher accuracy than a toposheet. The morphometric parameter is heavily depended on the scale of the feature extracted. Research is going on which satellite imagery is suitable for accurately extracting the drainage network in various surface areas like mountain, flat surface, less complex terrain area. In some cases, SRTM is very good for drainage analysis [15, 16]. For accuracy assessment, the absolute elevation parameters are more focused and less focus is given to how the various morphometric variables that are derived vary from one dataset to the other, as well as how their prepared maps differ.

3 Importance of Drainage Network in GIS

Drainage means the removal of excess water from a given place. In geomorphology and hydrology, a drainage network or river systems are the patterns formed by the streams, rivers, and lakes in a particular drainage basin. They depend on topography of the land. The number, size, and shape of the drainage basins found in an area vary as larger the topographic map [1]. A drainage system is of six different types, namely dendritic drainage, parallel drainage, trellis drainage, radial drainage, rectangular drainage and deranged drainage. In hydrological studies, geomorphological analysis DEMs are primary element for delineation of drainage network, catchment boundary, and estimation of various catchment parameters such as slope, contours, aspects and morphometric attribute like number of tributaries, stream length, stream order, bifurcation ratio, relief ratio. By examining various hydrological and morphological parameters, the irrigation management department supply water in dry weather for agriculture crop production, water resource management department, flood hazard zone prediction and mitigation department, drainage management department are being very beneficial in recent year.

4 Review of Drainage Extraction Methods

The mesh network of interconnected stream pertaining to a terrain is the river pattern or drainage pattern. These network formations mainly depend on the morphological aspect of the terrain, i.e., slope, varied resistance of rocks, and geology and topology of the land. When a DEM is considered for drainage network extraction, the main steps are (1) fill depression, (2) flow direction, (3) flow accumulation, and (4) stream network generation. A digital elevation model (DEM) is the representation of terrain elevations in digital form usually stored as a rectangular array with integer or floating-point values. Among various algorithms, the DEM pixel computation is based on D8 method that is first introduced by O'Callaghan and Mark [17]. But this algorithm has some deficiencies. As per D8 algorithm, single flow direction is calculated by comparing the elevation of its eight neighboring cells where the cells with higher elevations flow toward adjacent cells with lower elevation as the water flows from high elevation to low due to gravity. However, parallel flow line generation in flat areas is the restriction of formation of concentrated channel flow and is a vital limitation of D8 method. The most important problem with drainage network delineation using DEM is the presence of sinks; for flat area and depressions, it is difficult to set the ends of drainage network and the assignment of flow direction in individual cells. Thus, for accurate extraction, the sinks are removed from DEM. In 1988, a newly developed algorithm is introduced by Jensen and Domingue to eliminate all "sinks" prior to the assignment of flow directions by increasing the elevation of nodes within each depression to the level of the lowest node on the depression boundary. Next, a new procedure for the representation of flow directions and calculation of upslope areas using rectangular grid digital elevation models is introduced known as D-infinity where the flow direction is not restricted to check its eight adjacent cells [18]. Some research is also done based on multipath flow direction, but it is time-consuming, and more manual effort is required for calculation [19, 20]. To improve the existing method, a path-based method to determine the nondispersive drainage flow direction in grid-based DEM is introduced. It improved the D8 to some extent but fails to eliminate local level bias [21]. Over the past 20 years, many improved method based on routing flow through pits and flats has been introduced. The technology is developed, and the drainage is extracted based on heuristic information. A novel algorithm is presented by W. Yang and co-authors in 2010 based on heuristic information that accurately extracts the drainage network but fails to detect unrealistic parallel drainage lines, unreal drainage lines, and spurious terrain features and has a closer match with the existing pattern [22]. More recently, in 2012, Mr. Magalhaes has proposed a very simple and innovative approach where the DEM is considered as island and the outside water level raises step by step until the whole DEM is submerged. So gradually, it floods the cells of the DEM, next fills the depression, and spreads it on flat to flow toward a neighbor if that neighbor has a defined flow direction that does not point back to the tested cell. In this way, flow direction assignments grow iteratively into flat surfaces from areas. After that, flow direction is calculated and flow accumulation is generated, that is the final step of stream network generation or drainage network computation [23]. In 2013, a flooding algorithm is proposed by Antonio et al. to extract the drainage on flat surface and able to work on unprocessed DEMs avoiding the problems caused by pits and flats and can generate watercourses with a width greater than one cell and detects fluvial landforms like lakes, marshes, or river islands that are not directly handled by most previous solutions [24].

5 Literature Survey

Research is going on which satellite imagery provides the accurate result for drainage extraction. Digital elevation models (DEMs) provide us a digital representation of the continuous land surface. A new generation of elevation data is generated by remote sensing technologies (e.g., SRTM, ASTER-GDEM, Cartosat-1 DEM) and is freely available. High-resolution DEM provides accurate drainage extraction. During the past 20 years, many satellite imageries are used to extract the drainage pattern and are examined on various terrains like mountain area, medium

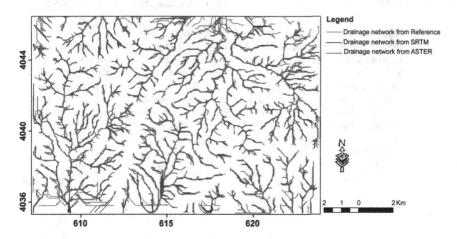

Fig. 1 Stream networks derived from Topo DEM, ASTER-GDEM, and SRTM [5]

complex terrain, and flat surface. But the accurate extraction which is very essential for hydrologic and morphological analysis is still lacking some features. K. Gajalakshmi and V. Anantharama recently analyze the accuracy between Cartosat-1 DEM and SRTM DEM. As per experiment in gradually undulating terrain, elevation values of Cartosat-DEM are lower than SRTM-DEM, whereas the stream parameter values of Cartosat-DEM are higher than SRTM-DEM [25]. Sarra Ouerghi et al. compare the ASTER-GDEM and SRTM DEM for drainage extraction. The analysis found that ASTER-GDEM is more pronounced in flat and less complex terrain [5]. Sample drainage network is shown in Fig. 1.

5.1 Summary of Survey

S. No.	Author and publication details	Title	Description	Comments
1	Paul Shane Frazier and Kenneth John Page, Photogrammetric Engineering Remote Sensing Vol. 66, No. 12, December 2000, pp. 1461–1467	Waterbody detection and delineation with Landsat 5 TM data	Landsat 5 TM+ imagery used to map river line waterbody and compared with aerial image	Manual classification of Landsat imagery and aerial imagery Error in image registration occurs
2	T. Toutin, Springer, INT. J. REMOTE SENSING, 20 NOVEMBER, 2002, VOL. 25, NO. 22, 5181–5193	DSM generation and evaluation from QuickBird stereo imagery with 3D physical modeling	QuickBird stereo imagery is used for generating the DEM and 5 m contour generation and drainage pattern extraction	Manual process to prepare DEM automatic extraction missing. Resolution is high but time-consuming process
3	Rajashree Vinod Bothale and A. K. Joshi and Y. V. N. Krishnamurthy, Springer, Indian Society of Remote Sensing, https://doi.org/10.1007/s12524-010-0258-8.2010	Cartosat-1 derived DEM (CartoDEM) toward parameter estimation of microwatersheds and comparison with ALTM DEM	Cartosat-1 and ALTM DEM is compared for drainage pattern extraction and microwatersheds parameter analysis in Madhya Pradesh	Suitable for mountain area but not suitable for plain area. Several steps required to process the DEM and extract the data

(continued)

(continued)

S. No.	Author and publication details	Title	Description	Comments
4	Samih B. Al Rawashdeh, Springer, https://doi.org/10.1007/s12517-012-0718-z,2012	Assessment of extraction drainage pattern from topographic maps based on photogrammetry	Aerial image and four toposheets are processed for drainage mapping and compare the both	Satellite imagery accurately extracts drainage, and some drainage is missing in toposheet. Several image processing steps are required to extract pattern from satellite image
5	Praveen Kumar Rai, Kshitij Mohan, Sameer Mishra, Aariz Ahmad, Varun Narayan Mishra, Springer, https://doi.org/10.1007/s13201-014-0238-y,2014	A GIS-based approach in drainage morphometric analysis of Kanhar River Basin, India	ASTER-DEM, Landsat ETM+, SOI toposheet for morphometric analysis	All morphometric parameters have not been introduced

6 Conclusion

The extraction of drainage networks can be done form of contour map, raster map, or DEMs. The satellite provided DEMs are very efficient to extract the drainage network in all terrain because the satellite gives the up-to-date changes that happen on earth surface like change in position of river network and new waterbody feature which are captured via highly sensitive sensor present in satellite. Thus, less time is required to extract the pattern from satellite provided DEM. Survey found that automatic extraction accurately extracts the pattern rather than topographic map with less error. The Cartosat-1 DEM is very useful for mountain area, and SRTM and ASTER-GDEM are suitable for medium and flat surface where terrain is less complex.

References

1. https://en.wikipedia.org/wiki/Drainage_system_(geomorphology)
2. Liu T, Miao Q, Xu P, Song J, Quan Y (2016) Color topographical map segmentation Algorithm based on linear element eatures. Springer J Multimedia Tools Appl 75(10):5417–5438
3. Al Rawashdeh SB (2013) Assessment of extraction drainage pattern from topographic maps based on photogrammetry. Springer Arab J Geosci 6(12):4873–4880

4. Toutin T (2004) DSM generation and evaluation from QuickBird stereo imagery with 3D physical modelling. Int J Remote Sens 25(22):5181–5193
5. Ouerghi S, ELsheikh RFA, Achour H, Bouazi S (2015) Evaluation and validation of recent freely-available ASTER-GDEM V.2, SRTM V.4.1 and the DEM derived from topographical map over SW Grombalia (test area) in North East of Tunisia. Springer Paper, J Geograph Inf Syst 7:266–279
6. Gajalakshmi K, Anantharama V (2015) Comparative study of Cartosat-DEM and SRTM-DEM on elevation data and terrain elements. Cloud Publ Int J Adv Remote Sens GIS 4(1):1361–1366
7. Toutin T, Chenier R, Carbonneau Y (2001) 3D geometric modelling of Ikonos Geo images. In: Proceedings of ISPRS joint workshop "High resolution from Space", Hannover
8. Toutin T (2002) DEM from stereo Landsat 7 ETM+ data over high relief areas. Int J Remote Sens 23(10):2133–2139
9. Poli D, Li Z, Gruen A (2002) SPOT-5/HRS stereo images orientation and automated DSM generation. Int Arch Photogram Remote Sens 35(B1):130–135
10. Hirano A, Welch R, Lang H (2003) Mapping from ASTER stereo image data: DEM validation and accuracy assessment. ISPRS J Photogram Remote Sens 57:356–370
11. Chopra R, Dhiman RD, Sharma PK (2005) Morphometric analysis of subwatersheds in Gurdaspur District Punjab using remote sensing and GIS techniques. J Indian Soc Remote Sens 33:531–539
12. Kale VS, Shejwalkar N (2007) Western Ghat escarpment evolution in the Deccan Basalt Province: geomorphic observations based on DEM analysis. J Geol Soc India 70:459–473
13. Sreedevi PD, Owais S, Khan HH, Ahmed S (2009) Morphometric analysis of a watershed of South India using SRTM Data and GIS. J Geol Soc India 73:543–552
14. Ghosh P, Sinha S, Misra A (2015) Morphometric properties of the trans-Himalayan river catchments: clues towards a relative chronology of orogenwide drainage integration. Geomorphology 233:127–141
15. Gorokhovich Y, Voustianiouk A (2006) Accuracy assessment of the processed-SRTM based elevation data by CGIAR using field data from USA and Thailand and its relation to the terrain characteristics. Remote Sens Environ 104:409–415
16. Weydahl DJ, Sagstuen J, Dick OB, Ronning H (2007) SRTM DEM accuracy over vegetated areas in Norway. Int J Remote Sens 28(16):3513–3527
17. O'Callaghan J, Mark DM (1984) The extraction of drainage networks from digital elevation data. Comput Vis Graph Image Process 28(3):323–344
18. Tarboron DG (1997) A new method for the determination of flow directions and upslope areas in grid digital elevation models. Water Resour Res 33(2):309–319
19. Zhang Y, Liu Y, Chen Z (2007) Multi-flow direction algorithms for extraction drainage network based on digital elevation model. Geospatial Inf Sci 6753(2B):1–9
20. Tarboton DG (1997) A new method for the determination of flow directions and upslope areas in grid digital elevation models. Water Resour Res 33(2):309–319
21. Orlandini S, Moretti G, Franchini M, Aldighieri B, Testa B (2003) Path-based methods for the determination of nondispersive drainage directions in grid-based digital elevation models. Water Resour Res 39(6):1144. https://doi.org/10.1029/2002wr001639
22. Yang W, Hou K, Yu F, Liu Z, Sun T (2010) A novel algorithm with heuristic information for extracting drainage networks from raster DEMs. Hydrol Earth Syst Sci Discuss 7:441–459
23. Magalhaes SVG, Andrade MVA, Franklin WR, Pena GC (2012) A new method for computing the drainage network based on raising the level of an ocean surrounding the terrain. In: Proceedings of 15th AGILE international conference on geographic information science, Avignon (France), pp 391–407
24. Rueda A, Noguera JM, Martínez-Cruz C (2013) A flooding algorithm for extracting drainage networks from unprocessed digital elevation models. Comput Geosci 59:116–123
25. Gajalakshmi K, Anantharama V (2015) Comparative study of Cartosat-DEM and SRTM-DEM on elevation data and terrain elements. Int J Adv Remote Sens GIS 2015 4 (1):1361–1366

Survey on Transmission Control Protocol Performance Over Different Mobile Ad Hoc Routing Protocols

Uttam Khawas and Kiran Gautam

Abstract Mobile ad hoc network is an infrastructure-less network where the nodes are mobile and each node behaves as a router. There are many routing protocols in MANET which are used to govern the path from the source node to destination node. Many problems are associated with the MANET due to its wireless nature and the dynamic topologies, so this survey paper focuses on behavior of Transmission Control Protocol in reactive and proactive routing protocols of MANET which are DSDV, AODV, and DSR.

Keywords MANET · Transmission Control Protocol · DSDV
DSR · AODV

1 Introduction

Mobile ad hoc network (MANET) is an autonomous mobile node forming a network which is infrastructure-less [1]. Each node behaves as a router and is independent of moving in and out of the temporary ad hoc network which gives the MANET a dynamic nature. As it is a wireless transmission, there are several reliability issues.

Since there are many nodes present in the network, the path finding is done using Wireless Ad Hoc Routing Protocols which are of many types. This paper focuses on some of the routing protocols which are of types:

(i) proactive and
(ii) reactive.

U. Khawas (✉) · K. Gautam
Sikkim Manipal Institute of Technology, Majitar, India
e-mail: Uttam.khawas@gmail.com

K. Gautam
e-mail: Kiran.gautam.cse@gmail.com

© Springer Nature Singapore Pte Ltd. 2019
H. K. D. Sarma et al. (eds.), *Advances in Communication,
Cloud, and Big Data*, Lecture Notes in Networks and Systems 31,
https://doi.org/10.1007/978-981-10-8911-4_17

Proactive are the routing protocols in which each node maintains tables containing the information about the network. The table is updated which is initiated by a certain node or done in a certain interval of time. Destination-Sequenced Distance Vector Routing (DSDV) is an example of proactive routing protocols.

Reactive Routing Protocols are routing protocols in which table are not present containing the information of the network, the path is built when the source node requires to transmit packet. It is the bandwidth-efficient protocols as the load for maintaining the table is not present. Ad hoc on-demand distance vector routing protocol is an example of reactive routing protocol.

After the path from the source node to the destination node is found, then the process of delivering the packet is done using the transport layer protocols.

One such protocol is Transmission Control Protocol which provides the reliability, error control, flow control, and delivery of packets in order. TCP is one of the most used Internet protocols and carries approximately 90% of Internet traffic [2]. TCP must be independent of the underlying networks [3]; i.e., it can be used for both wired and wireless networks, but it has been proven that TCP gives good results in wired networks but it does not apply same for the wireless ad hoc network.

2 Congestion Control Mechanism in TCP

Many data are lost in the network due to the congestion of the network, so the TCP performs the congestion control mechanisms.

1. Slow Start [2–4]

The sender starts the session with congestion window value of maximum segment size (MSS). It sends one MSS and waits for the acknowledgment, and after the acknowledgment is received within the retransmission time-out (RTO), the sender again sends two MSSs and waits for acknowledgment. This doubling effect after receiving each acknowledgment is known as slow start.

2. Congestion Avoidance

The doubling effect continues till it reaches the slow start threshold, and then, it goes linearly. This linear growth takes one MSS for each acknowledgment, and it continues till the sender congestion size reaches the receiver congestion size, which is known as the congestion avoidance.

3. Fast Retransmit [2–4]

If it does not receive the acknowledgment within the RTO, then it assumes that the packet is lost in the network. TCP may generate the duplicate acknowledgment when an out-of-order segment is received. If three or more duplicate ACKs are received in a row, it is a strong indication that the segment is lost. It performs the retransmission of the lost segment known as fast retransmit. After fast retransmit

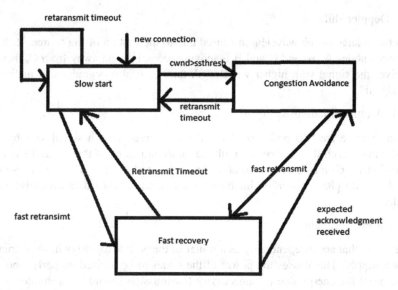

Fig. 1 Congestion control mechanism

sends the lost segment, then the congestion avoidance is performed without the slow start. This is the fast recovery process (Fig. 1).

3 Problems in MANET

(1) Frequent Path Breaks [2]

One of the main issues of the MANET is the node mobility and its velocity. Since the node is frequently moving and with a certain velocity, there are frequent path breaks. The path breaks make the TCP performance degrade and harder for the delivery of packets. It also complicates the routing protocols of MANET to find a path from the source node to the destination nodes.

(2) Lossy Channel [2, 3]

The errors in the wireless channels are caused due to:

(i) Signal attenuation

The strength of the signal reduces as it travels. Suppose P_s is the power of the transmitting source and P_d is the power of the receiving node and then:

$$P_s > P_d$$

It causes the quality of the signal to degrade with distance.

(ii) Doppler shift

It is the change of the wavelength caused due to the motion of the source. If there are three nodes A, B, and C and B is moving toward A and away from C, then A receives the signal with higher wavelength than C. It also degrades the quality of the signal.

(iii) Multipath fading

Suppose there are two nodes sender and the receiver, then a signal sent by the sender may reach the receiver by multiple paths and some of them may be due to reflection which may be received at a various interval of time. This will cause the problem with phase distortion intersymbol interference when data transmission is made.

(3) Power Constraint

The nodes that act independently as a router in the ad hoc topology have restricted power supply. This causes the power of the nodes to be utilized properly, and the wastage of the energy due to unnecessary transmission should be prohibited.

(4) Hidden and Exposed Node Problem

In hidden terminal, suppose there are three nodes A, B, and C as shown in Fig. 2, then a 'A' node tries to send the frame to B and at the same time C sends the frame to B and A is hidden from C, and there is a collision of frame at C [3].

In exposed terminal problem, suppose there are four nodes A, B, C, and D as shown in Fig. 3 and B is sending packet to A, the node C perceives as the channel is busy and does not send the packet to D. This causes the channel not to be utilized properly [3].

(5) Route Asymmetry

Since the nodes are mobile in the ad hoc topology, the path that is used for sending the packet and the path for sending back the acknowledgment may not be the same. Finding new path may be costly in terms of power consumption and delay in the delivery of packets [3].

Fig. 2 Hidden node problem

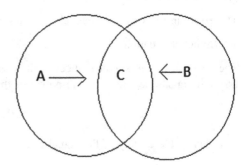

Fig. 3 Exposed node
problem

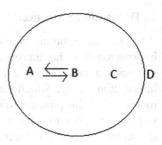

4 Performance Matrices

Some of the performance metrics that can be use to evaluate TCP performance in ad hoc routing protocols are:

1. Packet delivery ratio: It is the ratio of number of packets received by the receiver and the number of packets sent by the sender.
2. Average delay: It is the delay between the time from when the data packet is given to the IP layer at the source node and the received time of the data packet by the IP layer of the destination.
3. Throughput: It is the number of packets successfully transmitted to the final destination per unit time [2].
4. Packet drop: It is the number of packets dropped when the receiver buffer is full [2].

5 Routing Protocols

Wireless ad hoc network has many routing protocols which can be classified into four categories:

1. Routing information update mechanism,
2. Use of temporal information for routing,
3. Routing topology, and
4. Utilization of specific resources.

This paper focuses on routing protocols of the routing information update mechanisms, which are reactive and proactive routing protocols. So the routing protocols are as such:

1. Destination-sequenced distance vector routing protocols (proactive)
2. Dynamic source routing protocols (reactive)
3. Ad hoc on-demand distance vector routing protocol (reactive)

1. **Destination-Sequenced Distance Vector Routing Protocols (DSDV)**:

It is a proactive routing protocol which maintains the table containing the routing information. It is the improved version of the Bellman–Ford algorithm. Each node maintains a table which contains the shortest route and the neighboring node information through which we can reach that particular node. So the availability of routes makes this protocol to setup route with lesser delay. Table also contains the sequence number to remove the stale packets and the duplicate packets and to encounter the count to infinity problems. Tables are updated in a periodic manner or when a node observes that there is a significant amount of changes in the network [2, 5] (Fig. 4, Table 1).

2. **Dynamic Source Routing Protocol (DSR)**:

It is a reactive routing protocol which means that it does not have route table. Due to the lack of route table, periodic update of the table is not required which reduces the utilization of the bandwidth. Route is established when it is required, and it is done by sending the route request packet which is broadcasted in the network. The intermediate node forwards the packet by checking the sequence number of the packets received and before the time to leave has expired. When the destination node receives the route request, it sends back the route reply in the reverse order of the path from where the route request came with the information that contains the path traversed by route request [1, 2, 5] (Fig. 5).

3. **Ad Hoc On-Demand Distance Vector Routing Protocol (AODV)**:

It is a reactive routing protocol so route is established when required. The source node sends the route request packet to the intermediate nodes, and the source node and the intermediate node store the next hop information. The destination on receiving the route request sends back the route reply. The source node may get several route replies, so it uses the destination sequence number to get the up-to-date path to destination [1, 2, 5].

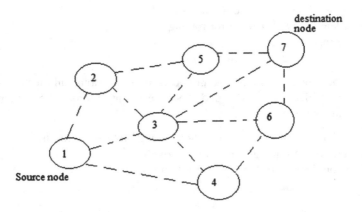

Fig. 4 Topology graph of the network

Table 1 Routing table for node 1

Destination	Next node	Distance	Sequence number
2	2	1	22
3	3	1	26
4	4	1	30
5	2	2	34
6	3	2	38
7	3	2	42

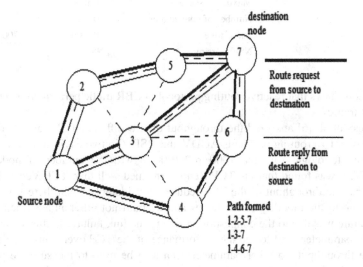

Fig. 5 Route establishment in DSR

6 Related Works

Noorani et al. [1] performed the analysis of two routing protocols, namely Ad Hoc On-Demand Distance Vector Routing (AODV) and Dynamic Source Routing (DSR) using the TCP Vegas with mobility consideration (Table 2).

The parameters used are packet delivery ratio and average end-to-end delay and found that AODV has highest packet drop and low average end-to-end delay using TCP Vegas. Furthermore, we can work on analysis of MANET environment under different issues such as node energy consumption, issues of hidden and exposed terminals, etc.

Chaudhary et al. [5] performed analysis of routing protocols such as AODV, DSR, DSDV under CBR and TCP traffic sources. The experimental results found that in CBR, if the traffic speed increases, the packet loss in DSDV goes higher than AODV and DSR. In TCP traffic, AODV has much higher packet drop than DSR and DSDV. So it shows that if the speed increases, load increases. The output of the

Table 2 Simulation
parameters [1]

Variables	Values
Simulation time	300 s
Topology size	1000 m * 800 m
Total nodes	50
Mobility model	Random way point
Traffic type	TCP vegas
Packet rate	4 packets/s
Packet size	512 bytes
Maximum speed	20 m/s
Number of connections	10
Pause time	10, 100, 250, 450, 700
NS-2 version	NS2-2.8

simulation shows that reactive routing protocol in CBR traffic performed better than in TCP traffic.

Tabesh et al. [6] analyzed the throughput of two different TCP variants (Reno, Vegas) over two routing protocols AODV and DSDV in two environment dynamic and static topologies of area 1000 m * 1000 m with 50 number of nodes for MANET. It was found that the TCP Reno performed well than TCP Vegas. It was found that the throughput of the TCP reduced considerably when there was a link failure due to the mobility of the nodes and it could not differentiate whether the node failure was due to the congestion or due to the link failure. In this simulation, the only parameter used to see the performance of the TCP over routing protocols was the throughput, so other parameters can also be used to measure the performance in the future works.

Jain et al. [7] performed the analysis of the three routing protocols AODV, DSR, and DSDV using the two traffic types TCP and CBR in a fixed map size with the pause time (0, 10, 20, 50, 100, 200). In this simulation, the results showed that the overall performance of the on-demand routing protocols was better.

Gururaj et al. [8] performed the comparison of the two TCP variants HSTCP and Reno in MANET environment. It was found that the congestion window drop rate is less in case of HSTCP when compared to Reno. Window size changes more dynamically and sharply in case of HSTCP and leading to larger window size.

Samit Rout et al. [9] performed analysis of TCP connection in mobile ad hoc network considering different network sizes of 70, 50, 30 in an area of 1000 m * 1000 m. The routing protocols that used were AODV, DSDV, and DSR. After the simulation results, it was found that throughput of the TCP increase slowly with increase of connection till it reaches 20 TCP connection. Packet loss of AODV was found highest, routing overhead of the DSDV was highest, and throughput of AODV was better than other two protocols.

Dr. (Mrs) Saylee Gharge et al. [2] performed the analysis of TCP variants which are Tahoe, Reno, New Reno, Vegas, Westwood, WestwoodNR, SACK, and Fack in two scenarios: (i) wireless link failure were 6 mobile node were considered

Table 3 Output in wireless link failure

Parameter	TCP variant giving best output	Protocol giving best output
Throughput	WestwoodNR	AODV
End-to-end delay	Vegas	DSDV
Packet drop	WestwoodNR/Fack	DSR

Table 4 Output in signal loss scenario

Parameter	TCP variant giving best output	Protocol giving best output
Throughput	All variants	DSR
End-to-end delay	Vegas	AODV
Packet drop	All variants	DSR

(ii) signal loss scenario were 3 mobile node were considered. Four routing protocols were used: AODV, DSDV, DSR, and AOMDV. Performance parameters used were throughput, delay, packet loss (Tables 3 and 4).

7 Conclusion

TCP performance deteriorates in MANET due to several reasons stated above. It does not perform as efficient as wired networks. The major issue is to differentiate between the packet loss due to congestion and the packet loss due to link failure due to mobility of the nodes in the MANET [6]. The routing protocols also effect the TCP performance as seen in the survey papers that in most cases on-demand routing protocol AODV gave better throughput than table-driven routing protocols [5, 9]. Packet loss is higher in AODV which is a reactive protocol than DSDV. We cannot justify which protocol performs the better results by taking few parameters such as throughput, packet, and loss delay. It was seen that protocol like AODV gave higher throughput but with greater packet loss [2, 9], so we cannot abruptly come into conclusion that certain protocol is best suited for TCP.

References

1. Noorani RN (2009) Comparative analysis of reactive MANET routing protocols under the traffic of TCP VEGAS with mobility considerations. In: 2009 international conference on emerging technologies (ICET 2009). IEEE
2. Gharge S, Valanjoo A (2014) Simulation based performance evaluation of TCP variants and routing protocols in Mobile Ad-hoc Networks. In: 2014 international conference on advances in engineering and technology research (ICAETR). IEEE

3. Al Hanbali A, Altman E, Nain P (2004) A survey of TCP over mobile ad hoc networks. Diss. Inria
4. Dyer TD, Boppana RV (2001) A comparison of TCP performance over three routing protocols for mobile ad hoc networks. In: Proceedings of the 2nd ACM international symposium on Mobile ad hoc networking & computing. ACM
5. Chaudhary MS, Singh V (2012) Simulation and analysis of routing protocol under CBR and TCP traffic source. In: 2012 international conference on communication systems and network technologies (CSNT). IEEE
6. Tabash IK, Ahmad N, Beg S (2010) Performance evaluation of TCP Reno and Vegas over different routing protocols for MANETs. In: 2010 IEEE 4th international symposium on advanced networks and telecommunication systems. IEEE
7. Jain R, Khairnar NB, Shrivastava L (2011) Comparitive study of three mobile ad-hoc network routing protocols under different traffic source. In: 2011 international conference on communication systems and network technologies (CSNT). IEEE
8. Gururaj HL, Ramesh B (2015) Congestion control for optimizing data transfer rate in mobile Ad-hoc networks using HSTCP. In: 2015 international conference on emerging research in electronics, computer science and technology (ICERECT). IEEE
9. Rout S et al (2014) Impact of multiple TCP connections in mobile ad-hoc network considering different network sizes. In: 2014 IEEE international conference on computational intelligence and computing research (ICCIC). IEEE

RETRACTED CHAPTER: A Survey on Detection and Mitigation of Distributed Denial-of-Service Attack in Named Data Networking

Sandesh Rai, Kalpana Sharma and Dependra Dhakal

The editors have retracted this article [1] because Figures 1 and 2 as well as parts of the text, were duplicated from a previously published article by the same authors [2]. In addition, Figure 3 was duplicated from a previously published article by Afanasyev et al [3]. This article is therefore redundant.

All authors agree to this retraction.

1. [Reference to https://link.springer.com/chapter/10.1007/978-981-10-8911-4_18]
2. [Reference to https://link.springer.com/chapter/10.1007/978-981-10-8237-5_51]
3. Afanasyev, P. Mahadevan, I. Moiseenko, E. Uzun and L. Zhang, "Interest flooding attack and countermeasures in Named Data Networking," 2013 IFIP Networking Conference, Brooklyn, NY, 2013, pp. 1–9.

© Springer Nature Singapore Pte Ltd. 2019, corrected publication 2020
H. K. D. Sarma et al. (eds.), *Advances in Communication,*
Cloud, and Big Data, Lecture Notes in Networks and Systems 31,
https://doi.org/10.1007/978-981-10-8911-4_18

RETRACTED CHAPTER

RETRACTED CHAPTER

RETRACTED CHAPTER

Author Index

© Springer Nature Singapore Pte Ltd. 2019
H. K. D. Sarma et al. (eds.), *Advances in Communication,
Cloud, and Big Data*, Lecture Notes in Networks and Systems 31,
https://doi.org/10.1007/978-981-10-8911-4

Printed in the United States
By Bookmasters

Printed in the United States
By Bookmasters